Restoring the Shining Waters

Restoring the Shining Waters

Superfund Success at Milltown, Montana

David Brooks

UNIVERSITY OF OKLAHOMA PRESS : NORMAN

Library of Congress Cataloging-in-Publication Data

Brooks, David, 1972–
Restoring the shining waters : Superfund success at Milltown, Montana /
David Brooks.
    pages cm
Includes bibliographical references and index.
ISBN 978-0-8061-4472-6 (hardcover : alk. paper)
1. Hazardous waste site remediation—Montana—Milltown.
2. Environmental policy—Montana—Milltown. 3. Milltown
(Mont.)—Environmental conditions. 4. United States. Comprehensive
Environmental Response, Compensation, and Liability Act of 1980.
I. Title.
TD1042.M92B76 2015
363.17'91—dc23   2014050155

The paper in this book meets the guidelines for permanence and durability
of the Committee on Production Guidelines for Book Longevity of the
Council on Library Resources, Inc. ∞

1 2 3 4 5 6 7 8 9 10

Contents

Illustrations

Preface

We were jobless and arrived in town on the last of what we had saved working overseas. Although I had lived in the West before heading to Japan, many things about Montana felt new and different, like the low density of people. Even in the city of Missoula, the houses seemed to float on seas of lawn grass, in contrast to the high-density apartments to which my wife and I had become accustomed. But what immediately surprised us the most was the smoke.

People who lived in western Montana during the summer of 2000 remember the smoke. My wife and I moved to Missoula near the end of that summer. As we drove from the eastern Montana city of Billings, over the Continental Divide, and 343 miles west to Missoula, almost all we saw were smoke-white horizons punctuated by the occasional blackened mountainside or fleck of fire. When we arrived in August, a digital photo of a pair of elk crossing the Bitterroot River just south of Missoula with the forests ablaze in the background had gone viral. Considered one of the most famous forest fire photos ever taken, that shot was one of many that captured blazes burning more than a million acres in the northern Rocky Mountains, with half that total in Montana, and nearly half the state's fires on the Bitterroot National Forest. As the smoke from those fires billowed north, it obscured the mountains ringing Missoula that normally made it a great place to hike, bike, run, or just be outdoors.

By August 2000, very few people were venturing outdoors in Missoula. The standing joke was that the air outside was so bad you were better off in a bar with a beer and a lung full of cigarette smoke than you would be in the woods. The latter wasn't an option, anyway. Federal, state, and local land management agencies had closed all public lands, including forests, open spaces, and rivers.

After a week in Missoula, we were sleeping on the floor of a college buddy's rental basement and had fallen in with a group of his friends who had grown up in the area. Having suffered a summer of smoke-induced confinement, they were starved for activity. One weekday afternoon, I piled into someone's pickup truck and headed up the Blackfoot River. At a riverside pullout, everyone except me grabbed his fly-fishing gear and disappeared up and down the river. Although I had never fished, my friend had wisely instructed me that the tool of the trade was a "rod" rather than a pole and that the Blackfoot had gained enormous fame with the release of Robert Redford's Hollywood version of *A River Runs through It* several years earlier. So, while they fished illegally I spent a few hours rambling along the riverbank looking at all the plants I didn't recognize, skipping rocks, and feeling pretty good about my new home.

During the drive back, I sipped a beer that someone had chilled in the river and listened to a few fish tales. As we neared town, talk turned to the issue of removing a dam that impounded the confluence of the Blackfoot with the Clark Fork a few miles upstream from Missoula. Having gained designation as a Superfund site (which I knew as much about as I knew about fly-fishing), the Milltown Dam was apparently drawing local attention because of a mounting environmental campaign to get rid of it. The idea of getting rid of a dam surprised me.

While teaching English in Japan, Becky and I had spent one of our spring breaks driving to the small island of Shikoku and canoeing for five days down the Shimanto River. It was the only undammed river in the country, according to our research. Even so, the retired physics professor who had rented us a boat picked us up on the last day under a highway overpass where the banks of the river were encased in concrete, like so many stretches of river I'd seen in the country. By contrast, the Blackfoot looked wild. I would not have guessed that a dam and a century of mining plagued it. Which is to say my knowledge of this place was as scant as what I knew about federal environmental law or catching trout with imitation bugs.

The prospect of a dam removal captured my interest because of what I did know, as well. The stories I had read about dam history in America were of them being built or of the efforts to prevent their building. They were stories about massive dams, exceptional dams. They did not include tales about the modest dams that are on nearly every appreciable stretch of moving water in this country, if not the world. And the dam stories I was familiar with had nothing to do with dismantling one. Hearing about Milltown Dam started me thinking about water, and rivers, from a new perspective.

My earliest memories are of swimming. They include diving down among the "weeds" at the bottom of a Florida Panhandle lake, and seeing how long I could hold my breath in a public swimming pool in a low-rent section of Tallahassee. Sometimes I revisit a fleeting memory that has me jumping off a dock into the icy water of Wakulla Springs, where the water was so clear you could see sixty-five feet down, or so I remember the driver of a glass-bottomed boat saying between pointing out alligators on the banks. My sisters and I floundered in the mud of what we had intended to be a backyard pool. After moving north from Florida to Indiana, I spent muggy summer days swimming in a small lake on the grounds of a monastery that was bordered on one side by one of the busiest streets in Indianapolis and by the city's largest river, the White, along another. In college I ran alongside and occasionally cross-country skied atop the Rio Grande as it glided south out of Colorado. As much as I sought out water for the enjoyment it gave me, I had never thought too hard about how people affected waterways. My day on the Blackfoot helped change that.

On the rare occasions I fly-fish, I spend more time untangling my line from the riverbank willows than I do spooking trout with my clunky casts. With the birth of our daughter a few years after moving to Missoula, Becky and I started spending a lot more time floating rivers. We've sought out a few with federal Wild and Scenic designation and also paddle down the Blackfoot a time or two each summer. By the time I started researching and writing about the history of the Milltown Dam in 2006, the Environmental Protection Agency had decided that dam removal and river restoration were the best solutions at the Milltown Superfund site.

The fate of Milltown, which was part of the largest such site in the country, seemed like more than a local story. Having continued to spend time on western Montana rivers and study the history of dams and water management in America, I thought that the convergence of a major federal environmental law, aggressive local advocacy, the removal of a dam, and restoration of the river section it had impounded was something new.

The process involved the efforts of lots of people acting on behalf of a river system they loved enough to break the law occasionally when they just had to fish. But they also loved it enough to spend thirty years participating in and changing a federal law designed to help clean up some of the most contaminated places in America.

Acknowledgments

I wrote this book in a basement room. Although the room is the smallest in my family's house and rarely accommodates more than one person sitting at a computer desk, I often felt like I was writing amid a crowd of supporters.

Reading and rereading old newspaper articles, EPA reports, and thousands of letters reminded me of both the writers of those documents and the archivists and librarians who helped me find them all. I kept a log at my side of all the thoughtful advice from many generous readers as I wrestled with the book's arguments, the organization of chapters, and the historical context of this story. Pen-marked pages of my draft chapters from many of those readers made it possible for me to reimagine the importance (or lack thereof) of historical evidence, rewrite each sentence, or rethink the choice of each word.

The first time I walked into Dan Flores's office to explain my interest in applying to the history PhD program at the University of Montana and my idea for a dissertation on the Milltown Superfund site, he offered encouragement and support for both endeavors. As my adviser, he guided my research with the combination of experience and curiosity that makes for fine teaching, and he also fostered my writing through example and sharp, necessary criticism. My other committee members contributed to this project in similar ways. Jeff Wiltse pushed me to connect the story of Milltown and Superfund with broader facets of twentieth-century American history. Tobin Shearer expanded my thinking on the history and functioning of environmentalism in the context of other social movements. As a Latin

American historian, Jody Pavilack reminded me to imagine audiences beyond environmental history or the regional. On the other hand, as experts in western literature and the history of the American West, Nancy Cook and Sara Dant were instrumental in my crafting a place-based narrative. I could not have asked for a more attentive, motivating, and supportive group of mentors.

I appreciate the candid and supportive help I always received from my fellow graduate students who offered early chapter reviews during our Rattlesnake Writing Group sessions. Shawn Bailey, Jeff Gailus, Monika Bilka, Greg Gordon, Ros Lapier, John Robinson, and Rachel Toor brought thoughtful criticism and an uncommon sense of fun to the process of historical research and writing. I am especially indebted to Greg and Rachel for allowing me to burden many fine runs, hikes, ski outings, and meals with conversation about my ideas and troubles with this project. Their input and example were invaluable.

I am also grateful for all of the people who took time to record oral histories about their involvement in the Milltown Superfund process for me. Although I have file cabinets full of government documents and newspaper stories about what happened at Milltown, personal stories, more than anything, helped me understand the truly important turning points and influences in this bit of history.

Tom Krause and Charles Rankin at the University of Oklahoma Press were instrumental in shepherding this book through the publishing process with a new author on their hands. Reviewers and copyeditors for the press helped guide my writing from raw manuscript to finished product. I received similar help from Molly Holz, who edited an article-length version of my story for *Montana: The Magazine of Western History*. Molly's photo editor, Glenda Bradshaw, helped me with her keen eye for images and her assistance in acquiring them. I owe my boss and friend, Gary Williams at Heritage Research Center, Inc., a special thanks (and probably a great deal more) for his unwavering faith in me and the flexibility he offered that allowed me to finish this book alongside my career.

My little basement office was silent during most of my writing. But the silence reminded me that my wife and daughter were off toiling at work and school. Being mindful of their efforts as well as their unflagging encouragement was a constant and welcome motivation to "git 'er done," as the saying in Montana goes. Above all, the playfulness and joy we indulge in as a family helped refresh me through all the challenges of completing this book. Any faults or omissions herein are mine alone.

Abbreviations and Acronyms

ARCO	Atlantic Richfield Company
ATSDR	Agency for Toxic Substances and Disease Registry
BDG	Bonner Development Group
CERCLA	Comprehensive Environmental Response, Compensation, and Liability Act, 1980
CFC	Clark Fork Coalition
CFRTAC	Clark Fork River Technical Assistance Committee
CFS	Combined Feasibility Study
CSKT	Confederated Salish and Kootenai Tribes
DHES	Department of Health and Environmental Sciences (also MDHES; see below)
DNRC	Department of Natural Resources and Conservation (also MDNRC; see below)
EPA	Environmental Protection Agency
ESA	Endangered Species Act, 1973
FERC	Federal Energy Regulatory Commission
FFS	Focused Feasibility Study
FS	feasibility study
FWP	Fish, Wildlife and Parks
MCCHD	Missoula City-County Health Department

MDHES   Montana Department of Health and Environmental
          Sciences
MDNRC   Montana Department of Natural Resources and
          Conservation
MEIC   Montana Environmental Information Center
MESS   Milltown EPA Superfund Site committee
MLWC   Missoula Light and Water Company
MontPIRG Montana Public Interest Research Group
MPC   Montana Power Company
MTAC   Montana Technical Advisory Committee

NEPA   National Environmental Policy Act, 1970
NPL   National Priorities List
NPRR   Northern Pacific Railroad
NRDLP   Natural Resource Damage Litigation Program
NRDP   Natural Resource Damage Program (same as NRDLP)

PRP   potentially responsible party

RI   remedial investigation

SARA   Superfund Amendments and Reauthorization Act, 1986
SDWA   Safe Drinking Water Act, 1974

TAC   technical advisory committee
TU   Trout Unlimited

UM   University of Montana
USFWS   U.S. Fish and Wildlife Service

Restoring the Shining Waters

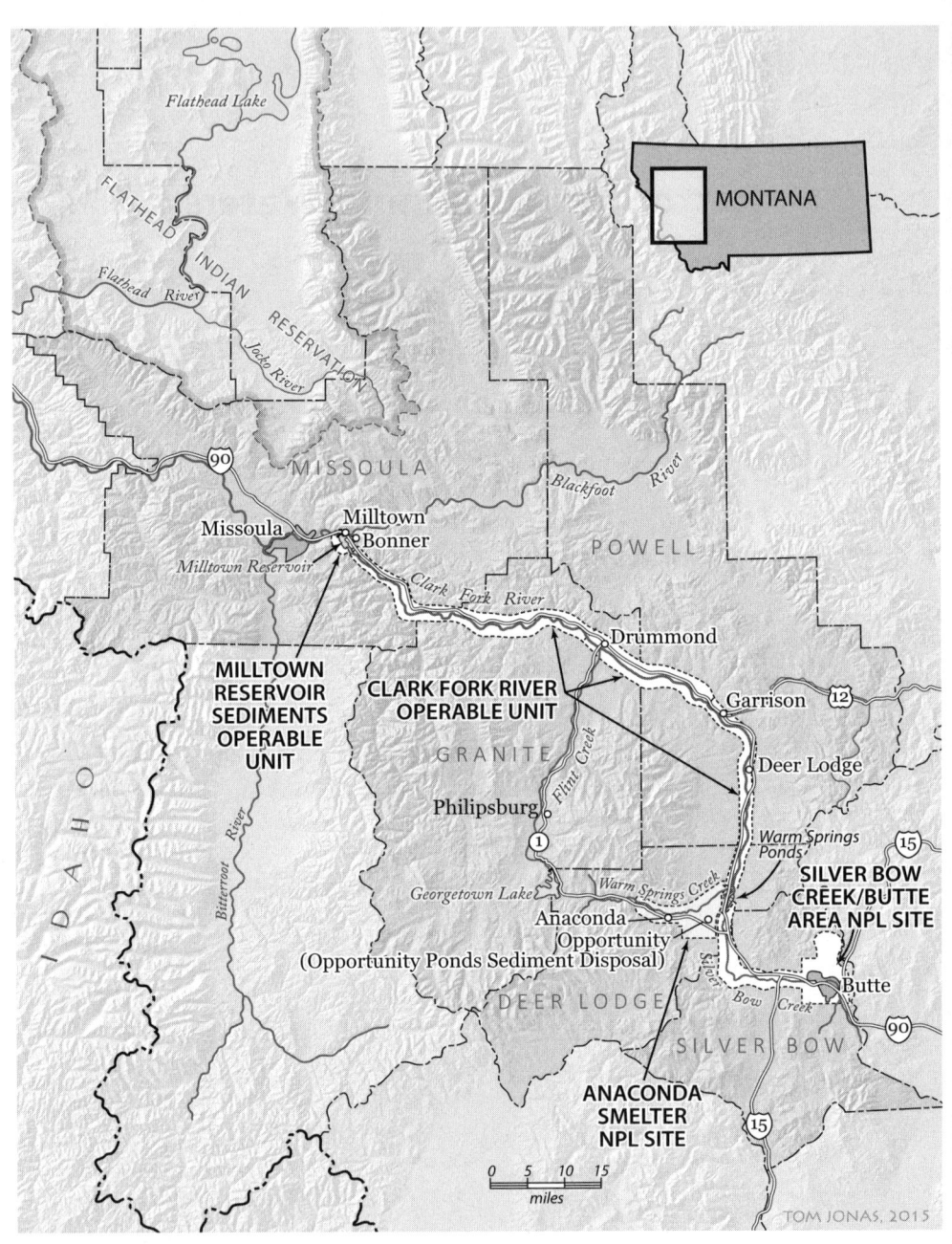

Milltown, Montana, and the Cark Fork River, 2014. *Map by Tom Jonas. Copyright © 2015 by the University of Oklahoma Press.*

Introduction

The notes of a single observer, even in a limited district, describing accurately its features, civil, natural and social, are of more interest, and often of more value, than the grander view and broader generalizations of history.

> —Anonymous epigraph to Bela Hubbard, *Memorials of a Half-Century* (1887)

The more a site feels like a place, the more fervently it is so cherished, the greater the potential concern at its violation or even the possibility of violation.

> —Lawrence Buell, *Writing for an Endangered World*

In 1980, the population of Milltown, Montana, was not enough to warrant a listing on the U.S. census, not even in combination with the nearby communities of Bonner and West Riverside. Most of the economic activity in those towns revolved around a plywood factory on the banks of the Blackfoot River, just upstream from its confluence with the Clark Fork. The heyday of milling in western Montana, and hence local population, had declined since the first decades of the century. That high point three-quarters of a century earlier had included the building of a modest hydroelectric dam that impounded the waters of the rivers' confluence to produce power for the local mills, businesses, and some residents.

Since 1908, the Milltown Dam and Reservoir had been fixtures of the community.

On the other side of the continent, Congress passed and President Jimmy Carter signed the Comprehensive Environmental Response, Compensation, and Liability Act (CERCLA) in 1980. Better known as Superfund because of the $1.6 billion cleanup trust fund the law established by taxing chemical and petroleum industries, it was meant to help identify and clean up the nation's worst toxic waste sites. Within a year of its passage, Superfund's first National Priorities List of such sites included Milltown. By 1983, the small western Montana community had joined a few hundred other places as home to the country's first federally designated Superfund sites. The Milltown Dam and Reservoir eventually became the tail end of the largest such site in America, encompassing roughly 120 miles of the Clark Fork watershed.

Superfund capped two decades of the nation's most vigorous environmental lawmaking. Precipitated by highly publicized environmental emergencies at places like Love Canal, New York, and Valley of the Drums (think oil barrels, rather than percussion instruments), Kentucky, the law aimed to stop, contain, and potentially clean up the dumping of toxic wastes by industries. Its namesake fund came from a tax on corporate dollars, and a fundamental stipulation of the law was that the "polluter pays." Where the Environmental Protection Agency, which oversaw the law's implementation, could connect a viable responsible party to a contaminated site, that party would have to pay for its remediation.

What happened at Milltown in the course of thirty years, from its designation as a Superfund site to its river restoration efforts, demonstrates some of the essential ways that Superfund changed throughout its implementation. The final remedy at Milltown was to remove an average-sized dam and restore sections of the Clark Fork ecosystem. After the EPA designated Milltown a national Superfund site, the environment itself, the persistent work of individuals within the channels of public policy, public comment, federally mediated compromise, corporate funding, a fish, and a Hollywood movie all shaped the Superfund process at Milltown. That process boosted the importance of public input in Superfund and of the EPA's increasing attention to restoring environments. At the same time, environmental groups such as those involved at Milltown began focusing on restoration of damaged environments. While Superfund cleanup at Milltown did push the law and environmental advocacy in new directions, many things about how it all happened were not exceptional.

Dams are found everywhere in America. In her book *Watershed: The Undamming of America*, Elizabeth Grossman pointed out that the Army Corps of Engineers could give only a rough count of "over 75,000 dams" on the country's waterways. While she summarized significant dam removal efforts in the late twentieth century, Grossman saw those endeavors as the beginning of restoring "approximately 600,000 miles of what were free-flowing rivers." Most of those river miles have lain stagnant as the dams that impede them have persisted for decades and, in many cases, for centuries. In other words, most people live near a dam at least something like the one at Milltown, Montana.[1]

As a Superfund site, Milltown was in many ways typical. The site had many of the significant features found at the "tens of thousands of hazardous waste sites" the EPA has located in the United States since passage of CERCLA. Attention to the site began because of human health concerns, which were the primary motivation behind the original Superfund bill and which continue to be a basis of site investigations and designations. At Milltown, arsenic contaminated the groundwater. According to a statistical analysis conducted for CERCLA, "25% of all Superfund sites have arsenic as the top contaminant." Groundwater contamination has been the most consistent and worrisome risk throughout the history of Superfund site designation.

Nearly a century of upstream mining caused Milltown's problem. Mining sites cost more and occur more frequently than any other type of Superfund cleanup. Two major industrial corporations were responsible for funding cleanup at Milltown. Nationally, nearly half of all Superfund sites have two to ten responsible parties. And, like Milltown, most sites that make the EPA's National Priorities List are in a rural-urban interface. The dam and reservoir's location in a cluster of small communities just upriver from one of Montana's largest cities and just downstream from expanses of relative wilderness made it representative within Superfund history. The dam itself was common. Like the great majority of the seventy-five thousand dams in America, Milltown was privately built and owned. Its modest size also made it representative of the kinds of dams on every river in the country except the Salmon in Idaho and the Yellowstone in Montana. The site also had some exceptional qualities that played into the law's implementation.[2]

~ ~ ~

Milltown developed as a timber-mining town at the confluence of two western rivers that are headwater streams of the Great River of the West, the Columbia. Both the Blackfoot and the Clark Fork gather from the melting snows and cold springs of the Continental Divide. The Blackfoot's icy waters form near a place where a small mine's thermometer once registered the lowest temperature in the continental United States at 69.7 degrees below zero. Norman Maclean, who grew up fishing and working in the woods of the Blackfoot drainage, wrote of this river in *A River Runs through It* that "it runs straight and hard." The river's upper two-thirds owes its swift flow, rocky bed, and steep, forested hillsides to the work of glaciers. Over the course of thousands of years, Ice Age glaciers in modern-day Washington created a reservoir that extended into the Blackfoot Valley. Dozens of times, those ice dams burst and unleashed the largest floods ever recorded on earth. The violence of those floods carved the Blackfoot's narrow lower third. So, lacking expansive floodplains and valley bottoms, the Blackfoot was a place of passage and short-term habitation to the first people who lived in the region. After millennia of use as an Indian route from the more temperate, open environments of western Montana to the bison-rich northern plains, the Blackfoot Valley became prime timber country to the area's first American industrialists. By 1908, the tens of thousands of logs the river floated from mountain camps to the sawmills of Bonner and Milltown backed up behind a new reservoir where the Blackfoot joins the Clark Fork.[3]

The Clark Fork arises on the flanks of the Continental Divide south of the Blackfoot. Unlike the Blackfoot, its water rushes off the mountainsides into streams that meander through broad valleys, a trait that earned the river's uppermost stretch the name Silver Bow Creek because of its long, looping course and its clarity. Similarly, Warm Springs Creek, the river's second major feeder stream, got its name because of the relatively warm water it contributes to the Clark Fork. The Clark Fork winds its way toward its meeting with the "hard and straight" running Blackfoot through 120 miles of willow and cottonwood bottomland that gives way to winter-hardy grasses on most of the valley floor. Or at least it once did.

The Blackfoot's timber industry grew up because of the state's mining boom. When the Milltown Dam was completed in 1908, it began supplying the power to mill millions of board feet of timber a year. Cut trees that were floated down the Blackfoot River helped build railroad tracks along the Clark Fork so the timber could reach the mines of Butte, Montana, supporting shafts as well as fueling smelters. The mountainsides that birth the Clark Fork are the work of massive igneous (volcanic) uplifting that formed a massive layer of granite. Molten gold, silver, copper, and other metals flowed and mineralized into the granite. When prospectors first began digging for riches in the area, however, they were also dredging up other, associated metals like cadmium, lead, and arsenic. Massive copper veins discovered in the 1870s and '80s in Butte prompted industrialization and consolidation of the area's mining by century's end. Company miners extracted tons of metal-filled ore out of the ground every day. Butte-born journalist Edwin Dobb reported in *Harper's Magazine* that "from the late 1800s through the first third of this [twentieth] century, the so-called Mining City yielded about 13.25 billion pounds of copper, which was a . . . sixth of the world supply. . . . Overall, the Hill has produced about 20 billion pounds of copper."[4] Whereas that copper rode the nation's first transcontinental railroad, the Northern Pacific (NPRR), along its path to helping electrify the country and then became valuable ammunition in both World Wars, the unwanted tailings, filled with toxic minerals, found their way into the Clark Fork through drifting, dumping, and flooding.[5]

In 1908, area residents celebrated the completion of Clark's Dam. Built by Butte mining magnate William A. Clark at the confluence of the Blackfoot River and the Clark Fork, the dam was meant to power sawmills and a streetcar line as well as electrify the growing city of Missoula, seven miles downstream. With far bigger banking and mining investments to oversee, Clark never bothered to visit his new dam. The locals who praised its contribution to western Montana's economy soon saw its benefits nearly washed away. The same year the Milltown Dam was completed, a record flood swept down the Clark Fork. Water that measured five times the average spring runoff bore millions of tons of mining waste downriver. The flood smothered the riversides with a mix of heavy metals. What didn't wash up on the banks or over the dam was trapped behind it. From there, the waste metals from Butte percolated into the surrounding groundwater. Seven miles downstream of the Milltown Dam, Missoula lost all its bridges over the Clark Fork in 1908 but celebrated the dam as saving the city from far more damage. Within a century, however, Missoula would be celebrating the removal of the Milltown Dam, which people had come to see as a threat rather than a savior.

Like that of Milltown, Missoula's growth was linked to Butte's. In 1883, when workers completed the NPRR by driving a proverbial "golden spike" into a wooden tie between Butte and Missoula at a place called Gold Creek, they quickened the link between western natural resources, like timber and precious minerals, and eastern markets. But not every tree, ingot of metal, or dollar rode the rails east. Along with industry, many urban centers developed in the West.[6]

Missoula began as a center of trade. Remembering a large valley where three rivers—the modern-day Bitterroot, Blackfoot, and Clark Fork—converged, a pair of railroad surveyors-turned-entrepreneurs returned to that valley in 1860. C. P. Higgins and Frank Worden opened a market and called their settlement Hellgate Village. The name referred to the narrow defile out of which the Clark Fork spilled into the valley, long known as a place where Plains Indians such as the Blackfeet raided western tribes like the Salish for their horses. Intertribal battles littered the dark canyon with the bones of fallen mounts and riders alike. Because of Hellgate Village's location at the hub of trade routes that people had been using for millennia, and because of its relatively mild climate, it prospered and grew. Within two decades the town included flour mills and sawmills. Bakeries, a brewery, fruit orchards, and an increasing complement of merchants and residents soon followed. Many of the things made or traded in Missoula were sold upriver in Butte. Before the end of the century, Montana had its first university, in Missoula. The city's new name was an abbreviation (or more likely, a mispronunciation) of the Salish word *Il-mis-eul-etsch-em*, which meant "shining waters." The name referred to a Salish story about how Coyote once threw the ashes of some beautiful women into the valley's river, bringing the deceased back to life as a riverside aspen grove and making the river that watered the trees glisten. The river eventually bore the name Clark Fork after explorer William Clark of the Lewis and Clark Corps of Discovery expedition. Meriwether Lewis traveled up the river through the Missoula Valley on the Corps' trek back east from the Pacific coast in 1806.[7]

By the time the EPA designated Milltown a Superfund site, the Clark Fork had lost much of its shine. While industrial logging and mining were the culprits above the Milltown Dam, Missoula was hardly isolated from that history or the river's contamination. Power from the Milltown Dam electrified the city, lighting businesses and providing it with streetcar service. By 1909, a year after the dam's building, two transcontinental rail lines hemmed in the Clark Fork for much of its course through western Montana. In downtown Missoula, the NPRR paralleled the river's north side and the new Chicago, Milwaukee, St. Paul and Pacific Railroad flanked its south banks. As Missoula bloomed into the "Garden City," a nickname it borrowed from Chicago, it also emulated that midwestern metropolis by becoming one of the largest mercantile centers in the northern Rocky Mountain West. With the expansion of trade, services, and population in Missoula, from a university to major medical facilities, moderate-sized manufacturing businesses, and a military fort that became a World War II internment camp, the Clark Fork carried an increasing burden of sewage and trash. Residents riprapped its banks through town with boulders and the occasional wrecked automobile. As was the case most everywhere in industrial America, people saw the Clark Fork as a natural asset because of its ability to carry off waste.[8]

When it came to implementing Superfund at Milltown, modern Missoula turned out to be up to the challenge. Historians of the American West have argued that an influx of fairly wealthy people seeking picturesque real estate, natural playgrounds, and a thriving service economy in moderate-sized western cities like Missoula have driven the rise in environmental advocacy in such places. In *Imagining the Big Open*, Liza Nicholas, Elaine M. Bapis, and Thomas J. Harvey compiled essays reflecting that sort of demographic shift and its influences. In his book *Bobos in Paradise*, David Brooks dubbed thirty-somethings, who had come to Missoula and Bozeman, to Boulder, Colorado, and to "half of the towns in northern California, Oregon, and Washington State" in the late twentieth century, Bourgeois Bohemians, or Bobos. The new wealthy class, or modern bourgeoisie, had taken on liberal leanings left over from the Beats and midcentury counterculture, argued Brooks. The Bobos failed to find bliss along the banks of the Blackfoot River, Brooks explained, and their attraction to such places was superficial, like a fashion, a way for people with expendable income to flaunt their environmental as well as other liberal values without really challenging the capitalist system that allowed them to do so.[9]

On the surface, Missoula certainly fits the description of a Bobo paradise. The old downtown sports plenty of art galleries and home furnishing vendors. Businesspeople along with the city's wealth of university and medical employees favor fleece and hiking boots or river sandals, depending on the season. Fundraisers for the town's abundant nonprofit organizations draw sit-down dinner crowds of generous donors. Auctions at such functions often include art and cases of wine. But those are not the only dynamics underlying the environmental advocacy in Missoula that drove the Superfund cleanup at Milltown. And they are not wholly new dynamics, as Brooks and others have claimed.

Missoula has always been home to educated, middle- and upper-class residents who take on social activism readily. Shortly after the town's founding, its proverbial fathers—mostly small-time entrepreneurs—convened for a few weeks of vigilante justice that resulted in a handful of hangings to keep the place safe from a growing criminal element and secure for the town's growing business interests. The town lobbied to get the state's first university, which it designed on land south of the Clark Fork donated by a local businessman. Missoula's Jeannette Rankin became the first woman elected to the U.S. Congress in 1916 and the only federal legislator to vote against the country's entry into both World Wars. When Montana rewrote its state constitution in 1972, its guarantee to state citizens of a "clean and healthful environment," as well as other provisions, made that constitution one of the most progressive in the nation. Representatives from Missoula at the Constitutional Convention helped spearhead that effort. As home to more than twelve hundred nonprofit organizations and a population of fewer than seventy thousand in the early twenty-first century, Missoula is routinely considered one of the foremost places in the country for tax-exempt social organizations. Social activism has a long history in Missoula; environmentalism was an extension of that tradition.[10]

Missoula's role in pushing the Superfund implementation in nearby Milltown was not the product of a radically new western demographic. Environmentally concerned citizens came to Missoula (often to attend the university) and stayed because of their love of the place and their attraction to the environment, and it was they who became committed to working for conservation organizations, local government, the university, and small businesses. It was the Confederated Salish and Kootenai Tribes that used their 150-year-old treaty rights to participate in an environmental issue that drew broad public interest. It was people who cared about the local environment because of their direct contact with it through recreation, study, work, and everyday endeavors. The median income in Missoula does not make it a wealthy city. A standing joke about the low wages in Missoula compared to elsewhere is that they are a result of the "mountain tax." Most people *sacrifice* income to live in Missoula. And not everyone in town is either a devoted or armchair environmentalist. Trucks with anti–wolf restoration bumper stickers vie in popularity with Subarus sporting save-your-favorite-environmental-thing slogans. And the place has two Wal-Marts, plenty of fast food, and, like most of America, a majority of eligible citizens who don't vote or participate in civic matters.

Even so, Missoula has a core of residents who value its outdoor op-portunities and who act routinely to protect and, in the case of Milltown, restore them. As surely as spring brings buds to the nonnative maple trees and lilac shrubs lining many of the valley's residential neighborhoods, cars bloom bright-colored kayaks on their roof racks and trucks sprout trailers bearing rubber rafts. Missoula's citizens frequently and overwhelmingly approve ballot measures to maintain or increase the city's more than three thousand acres of open space. In 1980, the year Congress passed Super-fund, Missoulians approved a half-million-dollar open-space bond; within fifteen years, voters upped their contribution to that cause to $5 million. Forming the north flank of the Clark Fork between Missoula and Mill-town, 1,465-acre Mount Jumbo cost taxpayers $2 million to conserve. Sim-ilar hillsides, large wooded tracts of land, and accessible peaks surround the city and offer hundreds of miles of trails to the people who have paid to keep them intact and accessible. The state's 1985 stream access law al-lows the public to use any river or stream for recreation within its ordinary high-water mark, regardless of land ownership adjacent to the waterway. Local businesses sport logos with native fish, mountain vistas, river runs, and other such place-based images. "Moose Drool," "Trout Slayer," and "Dancing Trout" are three of the best-selling beers brewed in Missoula. These are just a few snapshots of the values people in Missoula brought to the Milltown Dam Superfund process. And those values helped push the process toward dam removal and river restoration.[11]

That push represents one of the most significant trends in American environmentalism. In 2007, the journal *Environmental History* published an essay by Jenny Price that charted the history of American environmen-talism from its roots in preservation efforts to its dispersion into myriad competing issues and interests in the last quarter of the twentieth century. Yet, in this essay, titled "Remaking American Environmentalism: On the Banks of the L.A. River," Price offered river restoration as the new icon around which twenty-first-century environmentalism might rally. Describ-ing public efforts to restore the Los Angeles River into a free-flowing body of water within a fifty-mile greenway from its former tomb of concrete, she provided a profile of the kinds of efforts taking shape across the country, albeit in smaller ways. Restoration efforts like those of Los Angeles and Milltown are, according to Price, part of "an eco-frenzy without prece-dent." In his 2012 book *Recovering a Lost River: Removing Dams, Rewil-ding Salmon, and Revitalizing Communities*, Steven Hawley argues that this frenzy should extend to the removal of four dams on the Snake River that are much larger than Milltown was. The mounting push to remove dams and restore rivers is joining a broader restorative strain in twenty-first-century American environmentalism. What is happening on the Los Angeles River complements what is going on across the West, with the re-introduction of wolves to places like Montana; cutthroat trout, bull trout, and salmon recovery efforts in numerous watersheds; and the designation of new prairie preserves. The EPA has edged Superfund implementation toward restoration because of Milltown and its place in the much bigger context of restoring American environments.[12]

Public policy analyses confirm restoration's place at the forefront of American environmental law, as well as the impact public participation has had on that trend. In *Dam Politics*, William Lowry called river restoration "a new era for American river policies." His study of river restoration policy and cases showed that an increasing number of river restoration efforts, especially dam removals, hinged largely on public "receptivity" to the project, which in turn was a function of physical realities at each site. In particular, people's receptiveness to and willingness to advocate for dam removal and river restoration correlated with the "age, inactivity, and level of hazard" associated with the site, along with scientific evidence of the benefits of proposed changes. At Milltown, all of these factors were at play. However, a close look at Superfund implementation at Milltown suggests that other factors also motivated people to act on behalf of dam removal and restoration. Those factors ranged from business interests and scientific curiosity to the pursuit of recreation, pure fascination with fish or frogs, and routine public service work. The common element connecting the wide variety of motivations people had for participating in the Superfund process at Milltown was proximity to the site, the river. Proximity to a place with plenty of naturally functioning parts made it possible to restore the historically broken ones by participating in and steering the application of federal environmental law. As Lowry put it, "river restorations are indeed the manifestation of a new era in how humans manage aquatic ecosystems, but river restorations also represent one part of the late-twentieth-century transformation of how humans interact with their natural world." Fixing nature has become a priority, as written in the story of Milltown and at large in the ways Superfund law has evolved.[13]

Other than a few evaluations of Superfund statistics, most of the historical literature on the law focuses on its creation. That analysis has successfully shown the vigor of national environmentalism and federal legislating of environmental regulation during the 1970s, which included the Clean Air Act and the establishment of the EPA in 1970, the Endangered Species Act in 1973, and the Safe Drinking Water Act the following year. In the case of the ESA, historians have also paid attention to how some of that legislation has changed throughout its implementation. For example, the ESA has enabled the protection of a far wider variety of species than its creators perhaps intended, and it has provided a legal tool for preserving ecosystems rather than just their individual members. Studying the implementation of Superfund at Milltown provides the opportunity to see how major provisions of the law shaped environmental cleanup and vice versa. The process at Milltown altered the Superfund law, increasing the importance of public participation as well as pushing toxic cleanup toward the broader goal of ecosystem restoration. Milltown's Superfund history offers an opportunity to analyze the workings of environmental law where "the rubber hits the road," as a regional EPA administrator once described its local implementation.[14]

The proverbial rubber that hit Milltown's road included Superfund's essential provisions and some of its most important revisions. In chapter 1, I show that Superfund's mandate to identify the nation's worst toxic messes came about at Milltown not because the EPA scoured the country for contaminated environments but because local residents pressed local government to do routine water-quality tests. The work of local university scientists contributed, as did the advocacy of a range of citizens at the local and state levels. This chapter documents the law's purpose of attending to human health issues more than to environmental health during its early years. It also shows how Superfund went beyond regulating contemporary toxic waste with its focus on historic sources of industrial pollution. Chapter 2 charts the ways that designating a single Superfund site at Milltown led to a much broader recognition of the whole Clark Fork watershed as an environment abused by the area's legacy of industrial natural resource extraction. Part of that shift resulted from the environmental studies that Superfund demanded as a step toward determining how to remove or contain contamination. In addition, this chapter demonstrates the ways that the river helped determine remediation at Milltown.

Whereas chapter 2 traces how the problems at Milltown alerted people to watershed-wide issues, the following chapter goes in the opposite direction as the focus of both locals and the EPA narrowed onto the Milltown Dam. One of Superfund's most important provisions requires the designation of liable parties responsible for environmental cleanup costs. When this happened at Milltown, both the corporation liable for in-stream contaminants and the owner of the Milltown Dam garnered liability designation. Along with denying responsibility, the dam's owner infused the Superfund process with its first major controversy over relicensing the impoundment. In the context of liability issues, the local community made its first impact on Superfund law by convincing the EPA to include greater public participation in the process. This episode foisted the dam into the Superfund decisions at Milltown and exemplified the heart of EPA authority over environmental cleanup—liability.

No major modern environmental story is complete without a lawsuit. Chapter 4 introduces the most important legal suit in the Milltown story, whereby the state sued the largest liable party for damages to natural resources. As the largest suit of its kind in Superfund and state history, this action was one key to linking Superfund cleanup, or remediation of contamination, to ecosystem restoration. In addition to exploring the actions of state agencies, this portion of the story features a growing awareness by the public of the past and potential value of their watershed.

Chapter 5 reiterates the importance of the environment in stories of how people manage it. Herein the Superfund process at Milltown demonstrates how a spring ice flood impacted Superfund implementation. Along with restorative efforts on the Blackfoot River, Robert Redford's production of the film *A River Runs through It,* and the importance of the bull trout as a threatened species, the flood expanded the scope of what needed to be cleaned up at Milltown, as well as the purpose of doing so. While human health remained a vital impetus to Superfund, endangered species and ecosystem health became more prominent issues driving remediation decisions at Milltown. The public attention to these events helped drive that decision finally toward the possibility of removing the century-old dam.

A bumper sticker bearing the phrase "Remove the Dam, Restore the River" promoted an advertising campaign by the grassroots environmental group associated most closely with the Milltown Superfund process. Chapter 6 tells the story of how that campaign and the flood of public input that followed swayed the EPA to pursue cleanup remedies that included tearing down a dam. This episode demonstrates how Milltown changed the way the EPA handled public input at Superfund sites and the weight the agency gave that input. At the same time, the campaign helped push the focus of Superfund from being containing or removing toxic wastes to restoring broken environments.

Restoring the Shining Waters ends with an account of some of the key differences between remediation, Superfund's original purpose, and restoration, its new direction. Chapter 7 features one of the most unique restoration projects that evolved out of the Superfund project at Milltown. That project shows how restoration extended far beyond the boundaries of the original toxic site and encompassed both ecosystem and cultural renewal. This entire history features analysis of the major economic considerations surrounding Superfund, from its tax on industrial corporations and their liability for cleaning up the environment to debates over congressional renewal and expansion of the bill. Chapter 7 addresses the ways that Superfund has helped foster growth in the environmental restoration industry and boost local economies by helping turn places associated with waste into spaces for outdoor recreation. Along with restoration, Milltown became a model for the ways the EPA is beginning to use Superfund to redevelop local economies, rather than just cordon off or carry away the contamination that plagues them.

~ ~ ~

What happened at Milltown between the discovery of arsenic in a few dozen people's drinking water in 1980, the removal of Milltown Dam in 2008, and the ongoing restoration work at the site provides a chance to rethink the implementation of federal environmental law. That sort of place-based inspection reveals instances of individuals and collective public participation altering federal law through its legislative amendment and its practical application. For example, the outpouring of support for dam removal and river restoration as a solution at Milltown led to that decision, pushed the EPA to consider public input earlier in its decision-making process, and added weight to such input within the agency.

Many dams had been demolished in the United States before Milltown. But none of them had come down as part of a Superfund site. The final decision for Milltown demonstrates Superfund's evolution from a law aimed at responding quickly to contain and remove (if possible) large sources of contamination affecting human health, to a long and complex process where remediation is a first step toward restoring portions of whole ecosystems and redeveloping less destructive economic opportunities in their vicinity.

In the more than three decades since Superfund became law, environmentalists, politicians, administrators, and the public have often had cause to criticize it as less than super. If its history as seen through the story of

Milltown seems an oddly positive contrast to critiques of the law's sluggishness or ineffectiveness, that is partly because people involved in the process at Milltown made it positive and remember it as such. As with any story about environmental issues, there were controversies. People who committed entire careers to the process found it tedious at times. But it was a process that worked. Having worked in the Missoula river conservation community throughout the entire Milltown Superfund process, Trout Unlimited director Bruce Farling admitted that the EPA's decision to force two "very powerful corporations" to fund a dam removal and river cleanup project costing them hundreds of millions of dollars was "really, really big." Mostly, it "tells other communities what the possibilities are," he emphasized. With our wealth of environmental histories about degradation and failure, it's worth having a few like Milltown's that provide models of relative success.[15]

CHAPTER I

Arsenic in Old Places

Milltown's Path to Superfund Designation

Expect poison from the standing water.
—William Blake

(I) Whenever (A) any hazardous substance is released or there is a substantial threat of such a release into the environment, or (B) there is a release or substantial threat of release into the environment of any pollutant or contaminant which may present an imminent and substantial danger to the public health or welfare, the President is authorized to act.
—Comprehensive Environmental Response, Compensation, and Liability Act (Superfund), Section 9604

William Woessner and Johnnie Moore rented a chain saw for twenty-five bucks. Then the two junior professors at the University of Montana recruited some graduate students and drove to the Milltown Reservoir a few miles up the Clark Fork from Missoula. As Woessner remembered it, the saw's fourteen-inch blade failed to cut deep enough into the ice, so they took turns bashing at the frozen surface with an iron bar. Once through, they plunged a grab sampler—reminiscent of a hinged pair of hand-sized excavator buckets—down the hole, scooped a load of reservoir-bottom muck, hoisted it up, sealed it in a plastic bag, and labeled it. Skiing across the groaning surface of the reservoir with their sled of gear in tow, they

In the early 1980s, the testing of "bad-smelling water" in Milltown led to the tiny community becoming a poster child of the federal Superfund law designed to clean up hazardous waste sites. Located in western Montana more than one hundred miles downstream from the mining city of Butte—the source of Milltown's groundwater contamination—Milltown and its subsequent cleanup redefined the meaning of Superfund site. The problem emanated from contaminated by-products of copper smelting that moved downriver and collected for more than three-quarters of a century in the reservoir behind Milltown Dam on the Clark Fork. *Courtesy Diana Hammer, photographer, U.S. Environmental Protection Agency, Helena, Montana.*

sawed and bashed and retrieved samples of its silt-covered bottom from four locations. As they went, wind-whipped flurries of snow erased their tracks, and the work kept them warm only so long.

"What in the heck have you guys been doing with this thing?" the man at the tool rental shop asked when they returned the saw with its chain encased in ice. Woessner and Moore explained that they taught in the university's Geosciences Department and that collecting sediment samples from under the frozen reservoir brought them out on that February afternoon in 1982. They were following a hunch that the sediments were the source of the arsenic recently discovered in Milltown's community water wells.[1]

~ ~ ~

Residents of Milltown knew their water had problems. For years they had watched the black stains on their sinks, tubs, and toilet bowls grow darker. Clothes laundered at home turned rust red. Cars and houses took on a faint yellowish hue when people washed them. Some grew accustomed to the metallic taste of their tap water without ever growing oblivious to the displeasure. Others had given up drinking the stuff by the time health department officials took samples in 1981. That was when Milltown resident Uuno Hill called the department and asked someone to come out and test his water "because it smelled bad."[2]

Water testing was supposed to be a standard procedure by then. Montana's implementation of the Safe Drinking Water Act (SDWA) in 1974 mandated that such tests occur every three years. The Missoula City-County Health Department had tested the Milltown well that served Hill and other area residents prior to 1981, but only for basic sanitation, not for its chemical properties. When the state laboratory found that the sample taken in May of that year contained levels of arsenic far exceeding federal standards, health department officials returned to Milltown and took seven more samples.[3]

Yet the results of those samples remained undisclosed. Ten days before Christmas, reporter Kevin Miller first broke the news that both state and local health department officials had been sitting on knowledge of arsenic-laced water from Milltown wells for months. An anonymous informant told Miller about the results from the spring inspection. The state lab in Montana, run by the Department of Health and Environmental Sciences, did not analyze the seven additional samples until August because of what the department's director called "a personnel problem." The state lab then delayed informing the state or city health agencies about the elevated levels of arsenic for another month. Adding to this pattern of delays, health officers at both the city and state levels decided to keep the results from the public until they better understood the problem. "I hate to tell people, 'Hey, you've got arsenic in your water and that's all we know,'" said state environmental health officer Joe Aldegarie in defense of his department's decision to withhold its findings.[4]

In the days following Miller's December news story, details about the state's mishandling of the inspection and the contamination itself became clearer. Aldegarie apologized for the delays and hinted at a shake-up in the management of the state's lab. He and Missoula health officials turned their attention to Milltown's water. Four of the seven Milltown water samples registered up to 370 micrograms per liter (µg/L) of arsenic, which is roughly equal to the sweet content of a gallon of pure water with one and a half *grains* of sugar added to it.[5] When the SDWA became federal policy in 1974, the U.S. Environmental Protection Agency (EPA) deemed 50 µg/L the temporary "maximum tolerance level" for arsenic, pending more conclusive research about its effects on human health. By 1981 the EPA was considering lowering the standard to 10 µg/L. Meanwhile, Milltown's wells were registering seven times the federal limit of 50 µg/L. How those levels of arsenic-contaminated water might affect the people drinking it remained, as Aldegarie implied, poorly understood.

Poor understanding bred caution and more than a little confusion. The day after the first story about arsenic in Milltown water appeared, local health officials delivered letters to thirty-three residences connected to the four contaminated wells, recommending that people quit drinking or cooking with the water in their homes. That day's local paper reported that research correlated skin cancer with significantly lower levels of arsenic than were present in the Milltown case. In contrast, the same paper also noted that a city-county health official recognized studies showing that it takes much higher levels of arsenic than Milltown water contained to cause acute health problems. A day later another round of letters strengthened the warning about not consuming water from contaminated wells. In a follow-up news story, Aldegarie called the second letter "a stronger statement because we can't take any chances." Local health officials warned that the levels of arsenic in Milltown water posed "long-range health hazards for residents." By day three of the story's unfolding, the newspaper's editorial page struck a more fearful tone, reporting that arsenic is "pretty deadly stuff." In addition to eating away at internal organs, the paper warned, arsenic could cause nerve damage, skin cancer, and a panoply of symptoms such as "vomiting, diarrhea, muscle pain, headache, weak pulse and, to cap the climax, coma and death." But, it concluded, since levels of arsenic in Milltown were lower than those known to induce such problems, "there seems to be no reason for panic." A week after the story got out, follow-up well tests produced a water sample that had ten times the federal limit on arsenic. Missoula health officials announced plans to test Milltown residents' hair and fingernails for long-term arsenic accumulation beginning the first of the New Year, 1982. Even with limited and sometimes conflicting information about arsenic at hand, health officials were erring on the side of caution. In the public perception, precise measurements and conflicting research on arsenic mattered less than arsenic's infamy: it had become synonymous with toxicity. And, the EPA hinted at getting involved at Milltown.[6]

At first, it seemed that locals were taking the bleak news in stride. Referring to the possible danger of drinking her water, one woman said, "My hair seems to be getting darker instead of whiter, so I don't know."[7] Joking aside, Milltown residents began grappling with the problem of finding safe drinking water, and they were not alone. Besides attracting the local university's scientific community, the issue induced local politicians as well as state and federal officials to take up the cause. As the political sphere of involvement grew, Milltown gradually became a national toxic landmark. Milltown residents had to come to terms with the idea that their environment, the place where they worked and their children played, was contaminated and potentially deadly.

Just as Milltown residents were aware of their daily domestic exposure to less than perfect water, they understood their industrial surroundings. For almost a century, the local lumber mills had embedded Milltown in the nexus of natural resource extraction that characterized the West. Rail traffic had stopped in Milltown since the completion of the Northern Pacific's transcontinental route in 1883. Throughout most of the twentieth century, one of the world's most productive copper mines operated about a hundred miles upriver from Milltown in Butte. For the decade prior to the arsenic discovery, Bonner—Milltown's next-door neighbor—was home to the region's largest plywood mill. People intuited that the water contamination stemmed from this industrial heritage, but they were unaware of the extent or the exact source of the problem. Solving those mysteries would take years.

And the solution would soon put Milltown on the EPA's National Priorities List (NPL), the queue to Superfund designation. In fact, by 1983 arsenic contamination of a few dozen Milltown wells had earned the town's reservoir a place on the federal government's first NPL and its subsequent designation as one of the first Superfund sites in America. What began as a local need for uncontaminated water became a national project to restore a major western watershed. The earliest advocates for what became the nation's largest Superfund site were not environmental groups but local residents acting on intimate health concerns and awareness of the industrial past of their surroundings.

With the number of Superfund sites growing every year, it is worth thinking about how Milltown's designation evolved at the local level. The story of Superfund designation at Milltown offers some unique perspectives on ways in which the law has played out. Congress passed the Comprehensive Environmental Response, Compensation, and Liability Act in 1980 in response to a social justice movement led by Lois Gibbs and the Love Canal Homeowners Association in New York. CERCLA capped two decades of the nation's most vigorous environmental lawmaking. Precipitated by highly publicized hazardous waste emergencies at places like Love Canal, New York, and Valley of the Drums, Kentucky, the law aimed to identify, contain, and potentially clean up hazardous waste sites. Specifically, as a citizen's guide to CERCLA put it, Congress established the law, which quickly gained the nickname "Superfund," to clean up "uncontrolled or abandoned hazardous waste sites." The story of Milltown demonstrated how that definition was stretched from the outset of the law's implementation to include hazardous waste repositories *and* places where waste collected unintentionally, such as behind a dam more than one hundred miles downstream from the origin of the waste.[8]

The image of a community battling to redress its unwitting exposure to toxic contamination by a malevolent corporation, as Love Canal exemplified, does not fit Milltown. The contamination at Milltown certainly resulted from a combination of the legacy of industrial extraction methods and natural systems. Superfund designation was less a fight than the outgrowth of a functioning municipal regulatory system (water sampling), public comment and concern working in the usual avenues of participatory government, and cooperation between the local, state, and federal government, the public, and private business. The curiosity, concern, and efforts of local citizens shaped implementation of a remedy at Milltown, especially toward restoration. Their involvement went beyond CERCLA procedures for evaluating and choosing a cleanup plan. As an analysis of the original law put it, those procedures were written such that the "definition of 'remedy or remediation' establishes a preference for on-site management of hazardous substances." Congress crafted Superfund to encourage containment of hazardous substances, keeping them in place. The law considered even moving people away from Superfund sites to be potentially more feasible than moving the contamination.[9]

In addition, Superfund at Milltown came about in the absence of any organized environmentalism such as the established national groups or radical fringe—the personality split of environmentalism in the 1980s. Environmentalism, even in its local, mostly cooperative, and practical form, appeared in Milltown only after the Superfund designation. In the beginning the situation unfolded mostly via routine governance and the actions of individuals within the community and within the agencies involved.[10]

~ ~ ~

Two rivers meet next to Milltown. Yet for most of the town's history the union of rivers was hidden.

In Milltown one engine of industry was a dam built in 1907 to impound the confluence of the Blackfoot River and the Clark Fork. One of the many new hydroelectric dams being built in the new century, it powered the region's largest lumber mills, including those of the Western Lumber Company (owned by the dam's builder, mining magnate William A. Clark) and the Blackfoot Milling Company (owned by the Anaconda Company) in the town of Bonner less than a mile east of Milltown. The dam's generators also powered a streetcar line that ran between Milltown and Missoula, seven miles downstream, and another that traveled up and down the Bitterroot Valley, south of Missoula. Electricity from the dam lit Missoula's main streets and well-to-do homes and businesses.[11]

In addition to electrifying the region, the dam and the milling that it powered brought people to Milltown, most of whom were young men in search of work. With Milltown lumber, railroad crews tied the tracks that connected the forests to the mills, the mills to the mines, the mines to the smelters, the smelters to wire manufacturers, and the manufacturers to retailers. Builders used Milltown lumber to construct homes and businesses in Montana and wired them with Butte copper. In 1920, when the three Dufresne brothers moved to Milltown to work in the mills, the Anaconda Company, which owned the Bonner mill, had more than doubled the previous year's production, and the value of its lumber nearly equaled that of all other manufacturing in the county.[12]

Those who came for jobs were following a westward movement of the lumber industry. Many French Canadians and Norwegians left logged-out areas of New Brunswick and the Great Lakes for the prospect of work in western Montana's robust timber industry. Finns and Swedes tended to make their way to western Montana as first-generation immigrants who had lost their farmland, timberland, or artisanal businesses as industrialization and industrial-scale agriculture marched northward and eastward across Europe. Swedes tended to come as bachelors, the Norwegians and Finns as families that organized readily around social activities like saunas and churchgoing. For a few years before the establishment of the post office, in 1912, Milltown went by the name Finntown. By the time the Dufresnes arrived, World War I had dampened immigration. For the increasingly stable population in Milltown at that time, wages and work were good. Like many others, the Dufresne brothers used the opportunity to start families.[13]

Leo Dufresne was born in Milltown in 1923. He grew up in his parents' company house and became accustomed to the ethnically distinct bars and neighborhoods that defined the town. The summer after finishing high school, Leo started working at the Bonner mill. At that time, the company rented the modest clapboard houses around town for about twenty dollars a month. Even after Leo retired from mill work the company subsidized his rent. After Champion Lumber bought the mill from Anaconda in 1972, the company supplied workers with firewood for the winter. With the abatement of most labor issues following the World Wars, Leo Dufresne, like most of his fellow mill workers, appreciated his employer and the lifestyle afforded by working in the area. He especially enjoyed watching his sons grow up in the same environment he had.

Summers, Leo Dufresne's three boys swam in the Milltown Reservoir or in a favorite eddy on the Blackfoot River or Clark Fork that fed it, just as their father had in his youth. When the reservoir, which they commonly referred to as "the river" or "the lake," froze during the area's long winters, they ice skated. Autumn and spring offered the possibility of watching migrating trumpeter swans take respite in the calm water of the reservoir. Although those big white birds were too rare to hunt legally, ducks abounded and provided many families a taste of autumn's bounty. And except during the height of spring floods, young boys like the Dufresnes could and did fish year-round.[14]

Depending on the season, the Milltown Dam drew fishermen for different reasons. "In the fall, when they were spawning," Dufresne recalled, "they'd come up against the dam and everybody'd go down there and catch whitefish because they'd all be up against the dam." The northern reach of the Sapphire Mountains, which forms the southern edge of the Clark Fork's valley at Milltown, blocks sunlight from shining directly on the river and reservoir during the winter. So, the boys sat in the warmth of the powerhouse and dropped lines from its windows into the reservoir below when the dam's operator or engineer was kind enough to let them in from winter's cold. Many locals preferred catching the northern pike that populated the warmer, slower water in the reservoir as opposed to the trout inhabiting the cooler, swifter currents of the Clark Fork and Blackfoot. They enjoyed the pike's mild, oilier flesh more than that of the rivers' various trout species. Mill engineer John Price's son, along with his two cousins, once caught enough pike in just a few summer days to fill the family freezer. When his dad found the freezer full of roughly a hundred fish he told the boys they had to quit fishing until they ate all they had already caught. The boys soon convinced Mrs. Price to start cooking. They invited all the kids in town over for a fish fry and were back out fishing the next day.[15]

The harmony of living and working in Milltown did not blind people to its industrial character. People enjoyed the local environment, but they clearly knew that it was not native or natural. With generations of mill workers and lumbermen in their family histories, Milltown residents understood that the dam and reservoir were the foundation of a working environment on which they depended. In addition to the enjoyment that environment provided, it came with some annoyances. Even after the community drilled a few wells in the 1940s to service residences, some men hauled five-gallon buckets of water home from the mill at the end of their workday because the mill's well water tasted better and ran clearer than the water from the community well. Many of those men remembered seeing their parents walking between home and the company well with a yoke supporting buckets of water intended for laundering and bathing balanced on their shoulders. So, too, the local river water was not always pristine.

A short walk separated the two rivers that filled the reservoir, and the clear appearance of the Blackfoot often contrasted sharply with the plumes of reddish-brown water that coursed down the Clark Fork. Watching those plumes, Milltown residents understood the connection. How could they not? Wood products from their mills went upstream to Butte and Anaconda; industrial runoff returned along with the water that powered those mills. The rivers suffered from local customs as well. Some Milltown residents dumped their garbage and old junk off a local bridge into the Clark Fork. But no one saw even those instances of water pollution as poison. No one suspected the full extent to which the area's industrial legacy was affecting the environment.[16]

~ ~ ~

Arsenic contamination in Milltown wells hit the TV news immediately following the local paper's story in December 1981. After watching a report about the contamination, William Woessner dialed the Missoula City-County Health Department. Eager to get involved, Woessner introduced himself as a first-year professor at the University of Montana in Missoula with a background in hydrogeology. As a student of how water moves through the earth, he had published a few reports on relationships between groundwater and mining in the West. Health department officials were happy to enlist him in figuring out what was going on in Milltown. As he had hoped, Woessner soon found himself tracking down the source of the arsenic in Milltown wells.

Some Milltown residents pointed to local sources of contamination such as an old dump, the local plywood mill, the railroad tracks, or even the local habit of tossing everything from trash and old appliances to abandoned cars into the river. Woessner investigated the possibilities. The construction of Interstate 90 through town in the 1960s necessitated moving a makeshift dump. Construction crews used dirt and gravel from below the old dump for fill beneath some of the interstate footings near the reservoir. Woessner figured that since locals used the dump for household trash, it probably did not contain much, if any, arsenic. Some speculated that the town's plywood mill was the source of the arsenic. When the Anaconda Company liquidated its mills in 1972 in an attempt to keep its American mining operations afloat, U.S. Plywood bought the milling operations in Bonner and transformed them into one of the country's largest plywood manufacturers. Hundreds of Milltown residents who found jobs with the new wood products company after Anaconda laid them off pointed to their new employer's practice of dumping wood treatment chemicals along the banks of the Blackfoot River just before it reached the reservoir. Again, Woessner discounted that theory when he consulted mill records and found no historic use of arsenic at the site.

In the early reports that spurred Woessner's interest, state environmental health officer Joe Aldegarie speculated that "arsenic may have seeped into groundwater from the sludge behind the Milltown Dam on the Clark Fork River."[17] Local opinion offered the same culprit. Many residents pointed at those reddish-brown plumes coloring the Clark Fork. It was no secret that the Anaconda Company had produced tons of arsenic and other heavy metals for most of the twentieth century as by-products of the country's largest copper mining and smelting operation sitting at the headwaters of the Clark Fork. Those by-products had nowhere to go but with the flow of the river, and that flow slowed at Milltown. In addition, a graduate student in the university's Environmental Studies Program found exceptionally high levels of copper, zinc, iron, and manganese in the reservoir's sediment during her investigation of river-borne heavy metals in 1975. Although she did not test for it, Woessner understood that arsenic often occurs with other heavy metals and moves with them in the process of mining, milling, and smelting copper. Woessner felt confident he was closing in on the culprit. So he recruited his colleague Johnnie Moore, rented the chain saw, and headed to the frozen reservoir.[18]

The test results for the sediment Woessner and company retrieved from the bottom of the Milltown reservoir revealed that the samples had high levels of arsenic, increasing the suspicion that the reservoir was the source of the toxins contaminating the local groundwater. The question of how arsenic in the reservoir's sediments got into the aquifer that supplied Milltown's well water remained.[19]

~ ~ ~

Like mountain-fed rivers across the West, the Clark Fork reaches its peak flow in the spring. As days grow longer and temperatures rise, some of the year's mountain snowpack melts and courses downward. Rivers rise. As surface water wanes with warmer days and drier hillsides, river levels drop. Some of the mountain precipitation continues to feed the river by seeping downward and following ancient, water-carved paths through the earth. Lots of this groundwater eventually reaches alluvial aquifers. These water-bearing strata are made up of old riverbeds, so they are most often adjacent to, and slightly above, newer riverbeds. Because of their position and porosity, alluvial aquifers usually leak groundwater into the nearby surface water, often a river. This movement of groundwater to river characterizes most of the Clark Fork. At Milltown the reverse happens.

Woessner discovered that around Milltown, especially under the reservoir, surface water fed the groundwater. About a mile above the dam, the Clark Fork ceased falling its average ten feet per mile and bloated into the reservoir. Ten or twelve thousand years before the construction of Milltown Dam, another dam had drowned the area under nearly a thousand feet of water. Glacial Lake Missoula filled upward of forty times, over as many or more millennia. Glaciers flowed south, damming the ancient Clark Fork in modern-day Idaho. These ice dams backed up massive amounts of water. Each time, the weight of all that water and the effects of warming weather broke the ice dam and unleashed the greatest floods recorded on the earth's surface. Those floods poured into the Pacific, reshaping many miles of seafloor with a mix of earth, water, and ice. On their way to the ocean, the mountain-sized floods carved the Columbia River basin hundreds of feet deep in places. Closer to the original ice dams, the torrent gouged a moonlike surface into eastern Washington known as the scablands. The ice-encrusted waters of the glacial lakes terraced the hillsides around Missoula and Milltown with their abandoned shorelines. And where the Clark Fork's valley pinches into a narrow canyon around Milltown, those epic floods dropped a load of rocks in their wake. These "high-energy fluvial and catastrophic glacial-flood deposits" left a layer of boulders, sand, and silt up to ten meters thick in the stretch of valley around Milltown. The same narrow stretch of canyon that collected all that earthen debris became an ideal spot to build a dam. So, Milltown Reservoir sat atop that ancient layer of rubble, which leaked.[20]

In addition to figuring out that there was an anomalous downward gradient from the reservoir to the groundwater below, the University of Montana geologists gathered sediment samples and drilled wells in an attempt to quantify the amount of arsenic in the reservoir and assess its quality as a means of linking it to the contaminated wells. They mapped a wedge of sediments that thickened from half a meter three kilometers upstream from the dam to nearly eight meters at the dam site. The team's calculations showed that through all that sediment the reservoir discharged more than a million cubic feet of water per day into the adjacent groundwater system, or about a dozen Olympic-sized swimming pools worth of water. They suspected that the water that percolated through the contaminated Milltown sediments into the groundwater carried arsenic. Woessner soon had a chance to voice his suspicions.

This time, the Missoula City-County Health Department called Woessner. It wanted to know just how much arsenic-laced sediment the reservoir contained. Sitting in his office on a Sunday, Woessner reached up to an aerial photo of the Milltown Reservoir and used his thumb to estimate the reservoir's area. He then mentally multiplied the area by what he thought was the average depth of the sediments, roughly four feet. His "back of the envelope calculation" of 1,800 tons was "a lot of arsenic" spread throughout a potential volume of about 4.4 million tons of reservoir sediment. With all that arsenic sitting above the Milltown aquifer, it was hard for

Woessner to suppose there could be any other explanation for the contamination. Yet no one had proved that the arsenic in the wells and in the reservoir were of the same form, and hence connected. And at the time, the UM team lacked the funding to pursue such proof. For that, Woessner blamed national politics.[21]

Back in 1980, when the U.S. Congress established CERCLA, the act mandated that the EPA "locate, investigate, and clean up the worst sites nationwide."[22] Since the dawn of nuclear testing and the exposure of widespread pesticide use as human health threats, Americans had become increasingly concerned about pervasive industrial pollution and the toxins it introduced into their daily lives. The overwhelming bipartisan support for passage of the Clean Water Act in 1972, including a swift and sweeping congressional override of Nixon's effort to veto the bill, passage of the SDWA two years later, and passage of Superfund in 1980, demonstrated how the general concern over industrial pollution and environmental toxins had polarized around water more than around any other aspect of the environment.[23] Yet the incoming Reagan administration staunchly opposed implementing Superfund, which required the federal government to spend money on environmental remediation or pursue lawsuits against corporations responsible for polluted areas. Woessner had received a small amount of federal funding for his team's groundwater investigation at Milltown. But as part of Reagan's agenda the administration identified itself with states' rights, free markets, and thinly veiled antienvironmentalism. That agenda pushed the EPA to avoid litigation and seek to cooperate with responsible corporate parties to alleviate environmental problems in conjunction with states' wishes. At Milltown, since there was no definitive source of the problem, much less a legally responsible party, this simply meant no more EPA funding.[24] Nonetheless, a few months after his conversation with the Missoula health department's director, Woessner heard that the EPA had included Milltown Reservoir as part of its inaugural designation of the nation's most toxic places. The agency cited Woessner and Moore's initial conclusions in its decision.[25]

~ ~ ~

Getting a commitment from the EPA took more than Woessner's thumb on a map indicating that there was lots of arsenic around Milltown. While he was figuring out the science of the problem, others had begun responding in different ways. Local pressure did as much for the cause as did the mounting evidence.

Milltown residents organized a water users association in the spring of 1982. Already bearing the cost of hauling clean water to their homes, residents wrote twenty-dollar checks to set up the new association, hire a lawyer, and begin finding a new source of water, a task estimated to cost between $40,000 and $100,000. In the coming months, Champion International Corporation, owner of the local sawmill and the housing affected by the poisoned wells, volunteered to pay the new water users association's legal fees. The company also loaned the water users group a truck to haul uncontaminated water to their homes twice a week. A year after the water users association formed, the company gave the group the twelve acres of land on which the company houses with contaminated water stood. Outright ownership of that land and those homes would allow individuals or the group to apply for bank loans to help finance a new water supply or make much-needed home improvements. Buoyed by the news of the $120,000 gift of land, the association's president, Melody Fuch, had no problem with the deal's caveat that mineral rights still belonged to the Anaconda Company. "They can dig for all the arsenic they want," she gibed.[26]

Fuch and others welcomed additional forms of help. Montana People's Action, a social and environmental advocacy group, organized a letter-writing campaign aimed at getting Milltown on the EPA's National Priorities List for Superfund. Others showed concern about corporate liability for the arsenic seeping under Milltown. Bob Ream, a professor of wildlife management in the University of Montana's School of Forestry and the Democratic candidate for the state House seat covering the Missoula and Milltown voting districts, publically pressured the Atlantic Richfield Company (ARCO), owner of all the upstream mining operations formerly run by Anaconda, to pay for a thorough study of the problem. With local and state pressure rising, along with mounting scientific evidence at hand, the EPA moved Milltown up its priority list.[27]

Of course, the environmental realities at Milltown helped its chances of making the NPL. The Superfund law's Hazardous Ranking System scored sites based on their "probability of release," "nature of material that might be released," and "potential targets." Milltown rated high on all three variables. Toxins were already being released into wells via groundwater, which accounted for the maximum-risk pathway in Superfund evaluation. The main culprit at Milltown, arsenic, topped the Superfund list of risky contaminants. And Superfund was designed to prioritize protection of people's health, which was the primary "target" at Milltown. With those variables, Milltown easily ascended the rankings.[28]

Being higher on the priority list made Milltown more likely to receive federal assistance. As decreed by Superfund legislation, that assistance would amount to 90 percent of all expenditures on the site, if the state paid the complement. The EPA's representatives in Helena, the state capital, had not released the new rankings. That did not stop one of Montana's U.S. senators, Democrat John Melcher, from prematurely announcing that the higher ranking "qualifies Milltown to be considered for EPA Superfund." Melcher was not alone in celebrating.

Missoula health officials lauded Milltown's rising rank among the nation's top toxic places because it carried the possibility of federal money and because it meant the vision of Milltown's problem had shifted. EPA agent Jim Dunn described how the agency broadened the scope of the problem at Milltown, garnering it a higher ranking, when "instead of saying the groundwater was contaminated, we looked at the contamination as a symptom of another problem." Missoula's newspaper cheered the change in perspective and acknowledged that it thought the other problems were the reservoir, the river, and its historical upstream use. In 1982, a year after Milltown's contaminated wells first made news, the EPA designated the site one of 418 official candidates for cleanup efforts based on human health concerns. The head of the Milltown Homeowners Association called the announcement "a nice little early Christmas present."[29]

Of course, there was nothing in that holiday gift so far, except a federal promise to begin considering remedies. In another example of how Superfund progressed through the usual workings of government and public policy, that gift began to fill. As the New Year approached and a new session of the state legislature convened, there was a flurry of local and state efforts directed at moving the Milltown Superfund site forward. Montana governor Ted Schwinden assigned a staff member to investigate the issue and to recommend solutions and sources of funding. Montana senator Max Baucus (D) wrote to the EPA's regional administrator requesting the "assistance" that Superfund would provide for Milltown residents. Efforts by Missoula's newspaper leaned in the same direction when, between Christmas and the first days of 1983, an editorial urged locals to write to the EPA encouraging the agency to include Milltown in its final cut. The editorial warned Missoulians of the bigger worry they might face if the arsenic-saturated sediments in Milltown washed into the Missoula aquifer.

Many downstream residents were already aware of the potential problem the dam's sediments posed. Since the late 1970s people from Missoula had protested periodic drawdowns of the reservoir because the practice killed fish. At the time, no one understood that arsenic was part of the sediment contamination, but people advocating stricter regulation of drawdowns knew that toxic heavy metals were part of the problem. A "reader comment" submitted to the newspaper early in the New Year warned of a similar potential disaster. With the discovery of arsenic, the dam and reservoir suddenly became a toxic time bomb to people outside the immediately contaminated area, and the EPA appeared to be the community's best bet for disarmament.[30]

People were beginning to frame the problem as bigger than just drinking water for thirty-three Milltown residences. Yet many of the letters sent to the EPA kept the attention on the urgency of the local problem. One fifth grader captured the essence of the letters' emphasis on the most vulnerable victims of the contamination when he wrote, "There are many old people and children. Please help them." Similarly, a Milltown mother offered the anecdote that when she mixed her well water into her baby's bottle "it curdled the milk." Senator Melcher used the word "families" a dozen times in two pages of his letter to the EPA. The intimate nature of the contamination, literally entering people's homes and bodies, filled those letters.

Ironically, the actual effects arsenic contamination had on Milltown residents remained virtually unaccounted for. The Montana Public Interest Research Group (MontPIRG), a statewide advocacy group for public interests, conducted its own health study of people living in the homes connected to the contaminated wells. It found that residents reported a high incidence of respiratory and skin problems, both of which medical studies correlated with chronic arsenic exposure. News sources never reported the MontPIRG study, nor did health officials or the EPA ever follow up on it or conduct a more thorough investigation, as MontPIRG recommended. Rather, involvement in the issue was spreading far beyond the confines of Milltown, just as people were beginning to see the problem as stretching far upriver and into the area's deeper industrial history, as well as into the future of downstream populations. The language of intimate effects combined with the possibility of downstream contamination motivated people to seek federal help and money.[31]

In the end, EPA funding hinged on the state. Even if the EPA put Milltown on its final NPL, it could not deliver any funding to the Milltown project until Montana approved a similar appropriation. State authority was needed to trigger federal participation. The Montana state legislature, in the midst of its biennial ninety-day session, began pushing a "mini-Superfund" bill sponsored by Bob Ream, who had defeated his Republican rival, a Milltown resident. The EPA's director of Montana operations reminded Montanans that his agency had moved Milltown into the top two hundred of the country's potential Superfund sites. He signaled that Milltown would probably make the final list, especially if the state had already committed its matching funds calculated at $220,000. Montana People's Action helped Milltown residents promote the bill allocating those funds. Such efforts paid off in a near-unanimous decision. Upon signing the "mini-Superfund" bill, Governor Schwinden pronounced that the appropriation law "sends a message to the federal government that we are serious about protecting our citizens from the consequences of hazardous waste." When Schwinden said "hazardous waste," he was concluding that the arsenic came from Milltown's industrial heritage. If there was still any uncertainty in the scientific community as to where the arsenic came from, there was no such doubt in popular perceptions.[32]

~ ~ ~

With the state's checkbook at the ready, the EPA committed to Milltown. In May 1983, before the spring floods swelling the Clark Fork and Blackfoot River abated, the state and the EPA agreed to fund a thorough study of the problem and the search for a permanent source of clean water in Milltown. By July the EPA had awarded $570,000 to the project. Finally, Milltown seemed to be on the path to some answers and solutions.[33]

Getting to this point had taken a long time. Two years had elapsed from the time the Missoula City-County Health Department sampled Milltown's water until the EPA and the state committed funds to the arsenic-contaminated environment. In that time, local and state agencies delayed informing the public of the test results. Once they did, information on the extent of the problem and the true health concerns remained murky. Lack of funding hampered the scientific community. And conservative aspects of federal Superfund legislation hampered the federal government from getting involved until the state committed to the project. Meanwhile, Milltown residents took on much of the burden of looking after their own health and creating whatever temporary solutions they could. They also watched as perceptions of their local environment changed. Where they once saw familiar and much-loved surroundings that held the stories of many generations, now the image of a poisoned place emerged. The water, in which many Milltown children swam like their parents and grandparents before them, grew into the ominous specter of one of the nation's most fouled environments. And fouling it was one of history's most infamous poisons.[34]

While the EPA's designation of Milltown as a Superfund site both condemned it and offered it hope, it would also begin the process of revealing the great ghost of western history. As many Milltown residents intuited, the economy that built their world had a darker, potentially deadly side carried quietly by the surrounding water. Since the Safe Drinking Water Act of 1974, the EPA's efforts to address the problem of toxins in America's water supply had been aimed at stopping industrial sources of harmful chemical compounds and heavy metals from ever entering the water. With Superfund designation at Milltown, the EPA committed itself not just to remedying ongoing and future sources of industrial water contamination but to redressing the past as well. All of which hinged, initially, on human health concerns—a new understanding of the river system as a toxic environment that carried a century's worth of industrial mining waste into the intimate spaces of people's homes and bodies. Rather than through fighting, Milltown received Superfund designation through a cooperative effort that included local grade-schoolers on up through the ranks of federal agencies. Clean drinking water was on its way. But tainted tap water was to become just the beginning, the proverbial headwaters initiating a much bigger stream of change, as it were. In the wake of Superfund designation at Milltown, the focus would soon broaden and the participants multiply.[35]

CHAPTER 2

Floods of Change

From Tainted Wells to Threatened Watershed

Much of the frontier is river, and rivers are meant to bring men together, not to keep them apart.

—J. B. Jackson, *Landscape in Sight: Looking at America*

The discovery of arsenic in Milltown's wells began when a local resident complained to the health department about the quality of his drinking water. University of Montana scientists identified millions of cubic feet of sediment accumulated behind the Milltown Dam as the source of that arsenic. Public concern, supported by Montana politicians, led the EPA to designate the Milltown Reservoir as one of the nation's first Superfund sites. But it was not locals, state or federal officials, or even the dam's corporate owners who instigated the next major change in store for the Milltown Dam. It was the river.

May is usually a pivotal month for the Clark Fork watershed. It is the wettest month, and it is the first month in which the air temperature remains above freezing for more than half its days. As rain falls and snow melts, western Montana rivers swell. In May, the Clark Fork more than doubles its flow along its upper stretches, including where it passes through Milltown. Those are the averages, anyway.[1]

In 1986, spring high water came early for much of the Clark Fork basin. Unseasonably warm temperatures in late February meant that rain, rather than snow, fell for days. Seven miles downstream from Milltown, in

Missoula, the local newspaper blamed the early-season flood conditions on "Sunday's three-tenths of an inch of rainfall and Monday's halter-top temperatures." Rain and melting snow coursed across frozen ground, clogged sewer drains with unraked autumn leaves and a winter's worth of debris, and flooded streets throughout town. One morning's front-page story featured a photo of a man standing in knee-deep water while he tried to unclog a street drain with a shovel.[2]

As residents of Missoula mopped flooded basements, stacked sand-bags, avoided flooded streets, and generally marveled at the ponds in their neighborhoods, the Clark Fork grew. On Monday, February 24, the river rose from two to six feet through Missoula, which was well below the area's flood stage of eleven feet. But because it was such an early-season flood those extra four feet of river carried lots of ice. As that ice floated into Mill-town Reservoir it piled up behind the old wooden dam.

By Tuesday, ice floes had smashed through the top of the Milltown Dam, sending an eight-foot wall of water and wooden debris over its face. While "the river rose noticeably after the noon collapse, floating miniature icebergs and huge sodden logs through Missoula," according to newspaper accounts, representatives of the dam's owner—Montana Power Company—and state Fish, Wildlife and Parks (FWP) officials remained unworried about downstream flooding. What worried watchers was the dam's safety and the toxic sediment contained behind it. FWP biologists began testing the river for heavy metals immediately following the dam breach.[3]

After nearly two decades of managing the dam's operations, Emmett "Smitty" Smith had shepherded the aging structure through plenty of seasonal high-water events. But never had one threatened to wash out the entire dam. As the ice and logs pounded against the mostly wooden dam, Smith chose what he regarded as "a beating rather than getting killed." He closed all the dam's gates so that the reservoir would rise and float the threatening debris over the dam rather than smash through it. Smith's choice to sacrifice most of the wooden gates and spillway, which the ice and log jams pummeled as they careened over the dam's face, probably saved the dam. Afterward he said of his own decision, "If we hadn't done that . . . we'd have lost the whole dam. It would have taken it right down to bedrock." While Smith's experience wagered injury over annihilation, the flood-damaged dam set in motion a more pressing debate about the structure's future, one that would ultimately take it "right down to bedrock."[4]

~ ~ ~

The condition of the Milltown Dam after 1983, especially following the rather minor flood of 1986, altered discussions about the long-term fate of the Milltown Reservoir. Newly formed environmental groups, state and federal officials, and the dam's owners had to confront the growing perception of the Milltown Reservoir as part of a river-wide pollution problem. Concern about the toxic sediment behind the dam expanded from being about local drinking water contamination to threatening the health of the entire downstream river basin. Focus on human health widened into river health. That expansion of concern, which mounted following the 1986 flood, also fit into a coalescing number of studies aimed at quantifying the health of the upper Clark Fork. Local and state research efforts dovetailed with the ongoing Superfund study of the river, even as the fate of Superfund itself hung in the balance of congressional debate. Milltown weighed in on that balance, making its first mark on Superfund legislation.

Furthermore, most studies and attention began to target fish, especially wild trout, as the indicator of overall river health. Public and governmental acknowledgment of basin-wide pollution problems also prompted the possibility of designating the entire upper Clark Fork as a series of Superfund sites. The floodwaters that catalyzed these changes regarding the Milltown Superfund site were not the first to affect the dam, nor would they be the last. Nevertheless, the water that poured over the dam in late winter of 1986 forced local environmental groups and citizens, as well as state, federal, and company agents, to tackle the issue of what to do with the aging structure because of the threat it posed downstream.

~ ~ ~

The 1986 flood was no great anomaly. Just five years earlier, a seasonally predictable May deluge had damaged the dam's spillway, and less than a decade prior to that, ice floes riding floodwaters had dinged the structure. By the 1970s and '80s, the dam's age had made it susceptible to the river's force, yet the worst damage sustained by the dam had occurred just six months after its completion.[5]

Building the Milltown Dam had meant reshaping the Clark Fork. When work began in the fall of 1905, the Clark-Montana Realty Company, owned by copper-mining magnate and Montana senator William Andrews Clark, had more than a hundred men on its payroll along with twenty-eight teams of horses excavating what was then known as the Hell Gate, or sometimes, the Missoula River. The men, mostly recent Scandinavian immigrants, along with French Canadians and Slavs, dug diversion ditches so that they could scour the river bottom of sediment, rocks, and "large boulders" in an effort to strike bedrock for the dam's footings. To locals it seemed like a massive undertaking. Although contemporary newspaper accounts touted it as a large, expensive, modern marvel, it was actually a fairly modest project for its time. Bigger dams had long since existed.[6]

At nearly 400 feet long and 274 feet thick, a dam built outside Cairo, Egypt, around 3000 B.C. would have dwarfed Clark's Dam in sheer girth. While the Sadd el-Kafara Dam was an anomaly of antiquity, by the late nineteenth century dam construction was entering its heyday around the world. So was the science of dam building.[7]

By the late 1800s, especially in the United States, experience and empiricism gave way to theoretical models and a rigorous study of hydrology. Experimental dam design and construction included sloped rather than vertical faces; thinner, curved structures; and the replacement of wood or rock with cement and steel. Large arched dams made of reinforced steel were proving their effectiveness in places like the San Bernardino Mountains of California, where the Bear Valley Dam stood by 1884. The most important advancement was the use of water turbines in dams to produce hydroelectric energy. Engineers favored the rapid fall and narrow canyons of rivers in the mountainous regions of the American West for the building of hydropower dams. By 1893 the Colorado River outside Austin, Texas, boasted a 65-foot-high, 1,300-foot-long masonry dam that generated 15,000 horsepower. Although a 1900 flood destroyed that dam, others of equal size and capacity were impounding western rivers as the new century opened and Clark's project in western Montana got under way.[8]

So when boosters made claims about the largeness of the Clark Dam, they were ignoring, or were ignorant of, reality. In a five-year period surrounding the dam's inception and completion, *Engineering News-Record*, the nation's leading publication on the development of everything from rivets to railroads, failed to mention the Clark Dam at all. Such journals favored stories about truly large or modern dams. Measuring about 220 feet long and 40 feet high, the dam on the Clark Fork was quite average. Rather than being a modern exception, the Clark Dam represented a fading generation of construction technology, albeit one that lacked the perils of experimentation. In many respects, it symbolized the tens of thousands of dams erected across American waterways before the turn of the century.

And its power capacity, while an emblem of newness, was modest. In sum, Clark's project was common and conservative. Yet the construction methods and materials employed by the engineers of Clark's Dam, while not cutting-edge, were critical to the structure's future.[9]

In profile, the finished dam looked like a squat pyramid set atop a low, rectangular base. The water face, or upstream side of the pyramid, sloped at about thirty degrees and was covered in rock, or riprap. The air face, or downstream side, sloped slightly more steeply until just above the base, where its angle softened, like a long run-out on a child's slide. Small, outhouse-sized compartments made of 10" × 10" wooden timbers called "cribs" composed the matrix of the whole thing. Men filled each compartment with rock and gravel, hauling some from the river excavation project and some from the blasting of the canyon walls on either side of the dam meant to create flat side abutments. The name "crib and rock" for this type of construction came from its interior design and the materials. The tightly fitted timbers, which swelled when wet, and the rock, gravel, and "non-porous earthfill," as well as a few sections of concrete poured on the upstream face, combined to make the dam watertight. All the rock within the cribs, and the "massive durable riprap" piled along the upstream and downstream sides of the base and the sloping upstream face—meant to direct the weight of the reservoir water down on the dam as much as it pushed downstream—counterbalanced the wood's flotation, or what engineers called "uplift."[10]

Of his many tasks, engineer Jerry Rourke was particularly intent on designing a structure that could withstand "30000 Sec. ft." of the river's flow. "The biggest record I have seen was only 24000 Sec. ft.," he wrote in regard to his design choice. The dam would soon see much bigger water than Rourke anticipated.[11]

Whereas high water interfered with the pace of construction in Rourke's first year as dam engineer, in the second year floodwaters threatened the structure itself. By the spring of 1907, Rourke's replacement, George Slack, was both overseeing the construction of the brick powerhouse and making repairs to parts of the crib and rock dam that winter ice floes and spring flooding had damaged. Slack blamed his predecessor's efforts, or lack thereof, for the repairs he oversaw. "The leaks in the concrete wall are due to quality of workmanship," Slack criticized. His evidence that men, not materials, were at fault included the contention that "the cement used was the 'Atlas' brand, one of the best grades on the market." By the following year, with the reconstruction work behind him, Slack showed unbounded confidence in the retooled dam.[12]

The dam's builders had built what they thought was more than a sturdy river impoundment. They intended the structure to serve several functions. Clark's Dam was emblematic of a new wave of hydroelectric projects in the United States, but it did more than produce horsepower. In addition to taking over construction of the powerhouse, George Slack spent the winter of 1907 trying to buy riverfront property, especially land adjacent to the soon-to-be reservoir. Given that it was the Clark-Montana Realty Company building the dam, the purpose of buying land along the river in the vicinity of the dam made sense. The engineer was prospecting on future population growth in the area and hence a booming local land market. In considering a new lease agreement for property near the reservoir, Slack assessed that the ranch was "much more valuable this year than last on account of the close proximity of the pond and its possibilities as a pleasure resort."[13]

The company was not the only one to see the new dam as a harbinger of growth and economic opportunity. At least one landowner in the area drew up plans for a new town site. Another individual began searching for the best spot to open a saloon. Revealing both his Progressive Era attitude about working-class alcohol consumption and one of Clark's primary intentions in building a power-producing dam, Slack thought the saloon idea "would not be very desirable in case you should build a saw mill here."

A new sawmill was Clark's goal for the dam's power. Soon after the company fired up its turbines in early 1908, Slack launched the next stage of the project. He located ground for a new sawmill that would be close to its power source as well as its artery of commerce, the Northern Pacific Railroad line. Tapping the river's power to enable industrial extraction of natural resources also entailed controlling life beyond the river. In particular, Slack engineered a spatial organization around the dam and new mill site. About this plan he inquired of his supervisors, "Are you going to have the people move their houses so as to fit the new lay-out of lots according to the drawings I sent you last summer? The arrangement is very irregular at present." Similarly, Slack sent his boss recommendations for the best location for a "pleasure resort or park." The dam was not only to control the river but to become the first piece in a new company town, which eschewed "irregular," vernacular development for "arrangement."[14]

Nonetheless, actual electrical power, rather than the abstract power to control space, was the primary purpose of the dam. On January 9, 1908, the testing of two of the facility's six turbines marked the first ceremony celebrating the dam's completion. The *Daily Missoulian* praised the dam as "one of the most substantial ever erected for power purposes" and delighted in its output when "the first electricity generated at the new power house flashed over the wires at 3:14 o'clock . . . and for a period of several minutes the big plant was brilliantly illuminated." Celebrants in western Montana seemed to think they had their own Times Square or White City. The newspaper regaled the details of the dam's powerhouse, which contained six turbines capable of generating five thousand horsepower when fully operational, and which "will be sufficient to supply the needs of the western portion of the state for many years to come." Once the dam's full power supply was connected to the mill's existing power plant, located nearby, it would more than double its capacity, which was already enough to electrify the local mills and every business and residence in Missoula into the near future.[15]

Those present at the ceremony had growth in mind when they contemplated this new power. When A. H. Wethey, superintendent of Clark's enterprises in Montana, acknowledged that the dam's final price tag of almost $400,000 was "nearly twice as much as we first anticipated," he showed no signs of regret about the expenditures. "No expense was spared," reported the *Daily Missoulian*, which gave an account of the tremendous quantity of resources used in erecting the dam, including "two million feet of timber . . . 5,000 barrels of cement," and too many thousands of tons of rock and granite to count—all of which was then backing up over a mile and a half of the Clark Fork and almost half as much of the Blackfoot to depths of up to twenty-seven feet just behind the new dam. All that water, held back by all that material, meant power—power for the future. Wethey imagined that the dam would "furnish power for local manufactories which are bound to locate in this section of the state," because "there seems to be no limit to the wonderful possibilities that can be accomplished with this great power." Since the inception of the dam, Wethey and Clark's company had had at least one possibility in mind.[16]

When the dam was still in the speculative stage, the first purpose of its power already seemed to be a new electric streetcar business. As early as 1904, the manager of the Missoula Light and Water Company (MLWC) was fielding letters about a prospective "Clark Damm" [*sic*] and a "street railway in Missoula" that it might power. By the time the prospect of a dam became a reality and construction started on September 13, 1905, Clark was in the process of acquiring the MLWC for $900,000. The deal meant Clark would provide power and water for nearly all public and private enterprises in and around Missoula, which already included manufacturers of beer, milled grain, and lumber. Diversifying his investments, including ownership of mines, banks, and public works, had been Clark's business strategy from his first entrepreneurial adventures in the Montana gold camps of the 1860s and '70s.[17]

Harnessing the river's power to help fuel the growth of what would become one of the largest mercantile centers in the northern Rockies meant altering that river. Just as the first step in constructing Clark's Dam was excavating the river bottom, each step entailed some change in the environment. Obtaining construction materials meant cutting trees from the mountainsides that descended into the Blackfoot and Clark Fork drainages. Men blasted, picked, shoveled, and hauled rock and granite from the river bottom, nearby quarries, and the Hell Gate Canyon walls, which they dynamited to create flat abutments between which the dam would nestle. Far less local pathways of natural resource extraction brought cement and steel to western Montana, including manufacturers in Milwaukee, Wisconsin; Dayton, Ohio; and Chicago, Illinois—from the latter of which Missoula borrowed its nickname, the Garden City, as well as its ambition to become a center of manufacturing, mercantilism, and trade. Closing the dam's gates drowned the confluence of the rivers, as well as six hundred acres of riparian habitat, agricultural lands, and low-lying timber. The alterations to the surrounding hillsides and river basin certainly affected the rivers' natural flood cycles, but those alterations did not abate the cycles. Blocking the river to produce power also meant battling its natural flood regime. That battle began almost as soon as the dam's floodgates closed.[18]

~ ~ ~

At the dam's public unveiling in January 1908, Slack crowed that "the dam will be in such condition that the highest waters ever known in this vicinity will not affect it in the least." He would soon be eating spring crow.[19]

In March, the dam handled its first bout of big water. Slack reported ice, "logs, trees, and drift wood slowly going over the dam" without causing any damage or disruption to its power production.[20] May tested the dam more seriously. With heavy rain in the valleys and snow falling in the mountains, western Montana's rivers rose. Showing his first signs of worry, Slack made daily examinations of the structure, reporting on the rising height of the water pounding down the spillway and eventually cresting the dam itself. The last day of the month, he recorded that "water rose so rapidly yesterday that we had to take down the flashboards at the south end of the dam. . . . At present the water is five feet and three inches on the crest, and still rising." The engineer set a team of men to dumping rock around the dam's edges for extra bracing. Men also dropped a few sticks of dynamite in the river below the dam to clear out a growing tangle of debris.[21]

As the wall of water washing over the dam's crest swelled to eight feet, it carried away a pier attached to the powerhouse, which was filling with river water. Slack ordered his men to continue throwing as much rock as they could haul at the problem. Still, the rain outpaced the men.

It rained more in May 1908 than in any month recorded in the state weather bureau's twenty-eight-year history. By the sixth of June, rain had fallen in the Missoula Valley for thirty-three days in a row. Across the state, rivers ran higher than anyone had ever seen. Raging water demolished nearly every bridge and roadway within the floodplains from Butte to Missoula. Even small streams, such as Rattlesnake Creek in Missoula, swelled to the point of ripping two-story houses off their foundations and floating them downstream, taking out trees, bridges, and other homes along the way, until the massive flotsam dumped into the debris- and mud-filled Clark Fork or lurched up into tangled piles above the banks. As the water rose and wreaked havoc, people living near waterways moved to higher ground. Businesses shut down as the focus of whole towns and cities along the Clark Fork became flood watching.[22]

Because most transportation and communication routes in western Montana, including the tracks of the Northern Pacific Railroad, followed low-lying valleys, the floodwaters stranded people. A summary of the damage along the length of the Clark Fork told the story of waters sweeping away wagon bridges, floating houses off their foundations, washing out railroad beds, snapping telegraph and telephone poles, breaching dams, and driving people to ever higher ground.[23]

Most news stories recognized Missoula as having "suffered the worst of a great calamity," since the collapse of its bridges spliced the town in two. Yet even as people congregated to marvel at the river's load of trees, dead livestock, rooftops, outhouses, and barges of splintered lumber, and to gaze across the muddy torrent at homes, businesses, family, or friends whom they had no hope of reaching, they worried most about unseen happenings upriver.[24]

On June 5 a telegram reached Missoula carrying the news that flooding had washed out a power-producing dam and left Butte in temporary darkness. Although the dam failure was not on the Clark Fork, speculation that the brand-new Clark's Dam was in danger of bursting rose along with the river. Even as the rains subsided and the sun made its first appearance in more than a month, the fate of that structure gripped downstream residents. At one o'clock in the morning, June 6, the river crested. Nearly thirteen feet of water flowed over the dam's spillway. As Missoula residents contemplated a rupture in the dam, their fear reached "the highest pitch of excitement and alarm, and the whole city has been in a state of unrest unparalleled in its history for any cause," according to the local newspaper. Those fears included a forty-foot-tall wall of water carrying millions of feet of timber that would careen toward the city if the dam gave way.[25]

As quickly as it had mounted, fear of a dam failure subsided. Despite the damage the flood had already done, joy spread when word came from upstream that the dam "had passed the crisis of its test by flood." Observers praised the dam's strength. One man who walked the seven miles upstream from Missoula said of the structure, "There is no more danger of the power dam going out than there is of the mountains washing down the river." Even though he could not see the dam for all the water rushing over its entire length, he observed "not a particle of a sign of weakness in the structure . . . I never saw anything stand up better than that dam." While the dam did withstand floodwaters of roughly forty-eight thousand cubic feet per second during the height of the flood (nearly twice the high water its engineers had anticipated), it did not emerge unscathed.[26]

Two days after the river crested the dam got a more thorough inspection. One engineer reported that the powerhouse had sustained minor damage. Turbulence created by floodwater pouring over the dam and down the spillway had eroded its foundation. Through the following summer and into the fall, W. A. Clark's company rebuilt substantial portions of the dam.[27]

~ ~ ~

As the dam aged through most of the twentieth century, it needed regular patching. Water leaked through the structure and seeped under it. With increasing regularity the Clark-Montana Realty Company, later the

Montana Power Company (MPC) after it bought the dam in 1929, filled cracks and plugged seepage by lowering the reservoir and pouring more concrete into the problem areas. In March 1933 the MPC hired a diver to inspect concrete patches the company had applied three years earlier, as well as assess the dam's overall stability from a fish-eye view. Diver W. W. Blaine found "a hole of some considerable size . . . in the soft natural rock bottom about 10 ft. in diameter and from 3 to 4 ft. in depth." According to the company, such degradation was common and "not considered important."[28] In the early 1970s, a thorough safety inspection of the dam reported eroding concrete patches and significant undercutting that "could ultimately lead to partial failure of the dam." Plus, decades of water eddying at the dam's toe had scoured the structure's foundation, which the inspectors warned could "seriously jeopardize the stability and safety of [the] structure." Yet the MPC continued to patch its aging dam.[29]

After the discovery of arsenic in 1981 led to Superfund designation, the structure's problems garnered much greater concern. When leaks sprang through the dam in the fall of 1983, environmental groups, state agencies, and the dam's owners began to take the situation more seriously. Members of the Westslope chapter of Trout Unlimited voiced worries about both the leaks and the drawdown of the reservoir necessary to fix those leaks, which flushed toxic sediment downstream. Fisheries biologists from the state FWP agreed. Both groups committed to monitoring downstream water quality and fish kills if a drawdown was necessary. They also began expressing interest in more than simply patching the dam. Trout Unlimited president Ray Prill suggested that it was "time people looked at the dam and came up with some long-term solutions. We don't want them [MPC] to put a Band-Aid on it." His suggestion resonated with river watchers at all levels.[30]

Soon after the news of the leaks became public in 1983, the Federal Energy Regulatory Commission (FERC) ordered the MPC to lower the reservoir and address the dam's problems. Before the end of the year the MPC lowered the reservoir more than seven feet without any significant downstream water-quality issues and began contemplating its options. The attention this incident garnered, compared to a long history of drawdowns and dam repairs, made it clear that Milltown Dam and the toxic sediment it withheld was becoming an issue of river-wide health.[31]

Lowering the reservoir and fixing a few leaks in a dam known to hold back enough toxic sediment to warrant Superfund designation prompted the MPC to consider the structure's long-term future, much like Prill had advised. During the middle of November 1983, as the company undertook emergency repairs, an MPC spokesperson admitted that "the cost of maintenance and abandonment is being studied over the long haul." The company's manager of hydroengineering promised that by the New Year the MPC would finish assessing what it would do with the dam, which suddenly seemed to include "abandonment." While the company did little to define what abandonment meant, MPC representatives confirmed that the toxic material behind the dam was crucial to their decision-making process.[32]

~ ~ ~

The concern about the state's first Superfund site spread both upstream and downstream. Just as FWP and environmental groups tried to safeguard fishery health below the dam, they began trying to ensure the same above it. While the worry downstream was the release of toxic metals from the reservoir, above the dam groups focused on flushing the river of its industrial legacy and restoring a viable fishery. FWP agent Larry Peterman said of the river's first 120 miles, "Twenty years ago there was nothing living in the upper Clark Fork. As late as 12 years ago, the upper Clark Fork was a dead stream from periodic spills of pollutants. Today, the Clark Fork is a reclaimed stream." Contrasting the river's barren past with its promising present was part of Peterman's strategy to support his agency's efforts to ensure minimum flows in many of the river's most important tributary streams, as well as in the main waterway. It was also an acknowledgment that the same heavy metals that polluted Milltown sediments lined pockets of the floodplain from Butte to the reservoir.[33]

As attention radiated upstream and downstream of Milltown, "study" became the buzzword for the Clark Fork. In a 1984 leap day article, the *Missoulian* reported that Montana's U.S. representative Pat Williams, a Missoula resident, "joined congressmen from Idaho and Washington . . . in calling for an Environmental Protection Agency study of pollution" in the river. While the request stemmed from a controversy over a Missoula-area pulp mill discharging its waste into the river, its intent was to address the health of the whole river. As Williams recognized, "there exists no baseline of information or a long-term plan for future use and protection of this vital resource." Williams's concern was in concert with that of state agencies, public opinion, and, to a degree, potential polluters. The attention to the Clark Fork as a whole watershed, as opposed to a series of isolated problems such as fighting a single dam or source point of pollution, was an early example of a soon-to-be national trend of communities supporting river-wide study and advocacy.[34]

Within a month of Representative Williams's initial efforts to get the EPA involved in a river-wide study, the state health department launched a two-year study of water quality on the Clark Fork. The study area covered nearly 225 river miles, with Milltown Reservoir right in the middle. According to a spokesperson for the Montana Environmental Information Center, the state's foremost environmental watchdog since its founding in the mid-1970s, the health department study, much like Representative Williams's efforts at the federal level, happened because "hundreds of people in Idaho and Montana took the time to get involved." Steve Pilcher, the chief of the department's water-quality office, also acknowledged that the study resulted from "strong" and "outraged" public reaction over pollution and the river's overall health. "It is good news, indeed, that we have a government in Montana that does listen to its people," applauded the *Missoulian*.[35]

Public outcry provoked the study but the public was not going to be alone in paying for it. Unlike many environmental pollution problems, such as those at Love Canal, where public outcry led to acrimonious battles of opinion that often divided communities along racial or socioeconomic lines, those at Milltown benefited from a Superfund process that encouraged public input and interagency cooperation.[36]

~ ~ ~

While it was public pressure, touched off by the leaky dam, that instigated the river-wide study, the funding came from both public and private sources. The health department estimated that monitoring thirty stations along the river, including highly suspect areas such as reservoirs and mill sites, for "chemical, physical and biological water-quality variables" would cost $100,000 a year. Pilcher hoped that Williams's call for EPA support would include funds to match those of the state and an undefined contribution from local industries.

By May 1984, as spring runoff swelled the Clark Fork, Montana governor Ted Schwinden stood near the river's headwaters in Butte and announced that the Anaconda Company had given him a $50,000 check to be followed by another $150,000 over the next three years to help fund research on the Clark Fork. The governor acknowledged that the money's purpose of coordinating the health department's study with that of the

EPA was appropriate "because past Anaconda actions have impacted the Clark Fork drainage." With Anaconda's contribution—meant to coordinate river studies—public opinion, state and federal agencies, and industry had coalesced around the issue of river health, an issue that included everything from sewage to pulp mill pollution but that emanated from the toxic legacy of mining epitomized by the Milltown Superfund site. That is, the Milltown Superfund site was becoming part of a more thorough concern about river health.[37]

Dealing with the river's past had become the principal problem for its future. Phrases like "future use" and "future protection" sprang up in the public conversation about the river.[38] Considering the river's long-term health hardly erased its past. In 1984, during the first summer of the new study, FWP officials found thousands of dead fish floating in stretches of the Clark Fork above Milltown. Eventually they identified the "mystery pollutant" as "tailings loaded with copper and other metals" that a severe thunderstorm had washed from old mine wastes and into the river. It was the same pathway of contamination that had plagued the river for a century.[39]

From the late 1800s through the early twentieth century, mine owners from individual placer miners to the Anaconda conglomerate built a chain of extraction that spanned much of Montana. Trees cut from western Montana and milled in places like Milltown made their way to Butte, underlying railroad tracks and supporting mineshafts. The first industrial smelting operations included heap roasting, which was done on beds of whole trees laid out over hundreds of square meters. Miners piled raw ore on these open roasting pits, set them on fire, and let them burn for days, weeks, and sometimes months. The long, slow burning within these heaps oxidized the copper-sulfide ore, driving off much of the sulfur as smoke. Workers then moved the roasted ore to smelting furnaces, where the material was heated to its melting temperature, effecting a further chemical separation of copper from impurities. Finally, they poured the refined metal—mostly copper—into molds and loaded it on trains headed for manufacturing facilities that spun the copper into wire. The commodity chain continued until copper wire found its way into homes and businesses across America, including Milltown, where the electricity from the local dam pulsed through that same metal conductor.

But commodities weren't the only things to come full circle to Milltown. Concentrating ore in Butte and Anaconda left behind massive piles

The dam at Milltown was built in 1907 at the confluence of the Clark Fork and the Blackfoot River to provide hydropower to run the region's largest lumber mills, including those of Western Lumber Company, owned by the dam's financier, copper magnate William A. Clark, known as one of Butte's Copper Kings. Massive amounts of lumber were required to shore up the tunnels in the underground mines in Butte. *Courtesy Montana Historical Society Research Center, Photograph Archives, Helena, Montana, PAC 81–27.1, "General View of Sawmill Plant at Milltown, Montana—Western Lumber Co."*

of tailings, waste material that was no longer worth processing but still included a cocktail of heavy metals. The roasting process also filled the skies with smoke so thick that people sometimes carried lanterns at midday to guide themselves down the city's busy streets. After local women's clubs pushed through new local ordinances that shut down the practice of heap roasting in Butte because the acrid, sulfurous smoke killed their flower and vegetable gardens, mine owners began roasting ores in furnaces and discharging smoke through chimneys. By that time, Marcus Daly had built a smelter for his Anaconda mine at the south end of Deer Lodge Valley, adjacent to his new smelter town, Anaconda. A decade later, Deer Lodge Valley ranchers took up the fight against the smelter smoke that corroded the lungs of their cows and horses, killing many of the beasts within a year of grazing on vegetation dusted with arsenic from the smoke.

The ranchers were unable to stop the smelting but the federal government began to take notice of the smoke damage. The U.S. Forest Service documented the damage sulfur in the smelter smoke was causing to trees on national forest lands south and west of the smelter. In 1910, the United States filed suit against the Anaconda Company, asking the federal court to enjoin the company from discharging smoke that was damaging federal property.[40]

Even as residents fought to move and then regulate smelter smoke, that fine particulate matter fell to the ground. Rain and time acting on gravity's mandate washed it into the surface water along with the water-soluble heavy metals in the mountains of tailings that defined portions of the landscape around the smelting works of Butte and Anaconda. In Butte, toxic metals flowed into Silver Bow Creek, the headwaters of the upper Clark Fork. Smoke that rose and settled in Deer Lodge Valley did so into Warm Springs Creek, the river's next major tributary. Mining and smelting contamination that was killing roses and tomatoes in Butte, as well as livestock and conifer forests around Deer Lodge, began decimating the river. The waste, the unwanted material left over from the smelting, was heaped alongside Butte's Silver Bow Creek. Filled with toxic minerals, the fine tailings found their way into the Clark Fork through drifting, dumping, and flooding.[41]

In the late nineteenth and early twentieth centuries, the "environmental damage caused by the smoke and tailings discharged from the smelters" of the Anaconda Company (AMC) grew to "world-class proportions" akin to its copper production. Historian Fred Quivik's exhaustive account of the company's concentrating and smelting operations documented one facility in Butte that handled an average of "335 tons of ore daily," out of which came "200 tons of tailings, which it discharged along Silver Bow Creek," considered an industrial sewer by the 1890s. A conservative estimate showed that Butte concentrating works, most owned by the AMC, dumped eight hundred tons of tailings per day into and along Silver Bow. Spring runoff and precipitation eroded the fine-grained tailings into and down the creek, destined for the Clark Fork.[42]

As the concentrating and smelting operations moved downstream to Anaconda, Montana, the same process of disposing waste along watercourses continued. As early as the 1890s, downstream residents and farmers complained of gravel turning green, fish dying, and crops withering due to the AMC discharging hundreds of tons of tailings into and along

The environmental damage caused by the smoke and tailings discharged from the copper smelters in Butte and nearby Anaconda grew to "world-class proportions" in the late nineteenth and early twentieth centuries. Tons of tailings were discharged along Silver Bow Creek, a Clark Fork tributary, and seasonal flows washed the waste downstream to accumulate behind Milltown Dam. In this undated photograph, Butte smelters fill the sky with smoke, obscuring most of the city. The Colorado Smelter in the foreground was in the southwest part of Butte, less than a mile west of Montana Street and south of what is now Interstate 90. The neighborhood next to it was called Williamsburg. *Courtesy Montana Historical Society Research Center, Photograph Archives, Helena, Montana, Lot 5 b5 f11.01, "Butte and Colorado Smelter."*

Warm Springs Creek every day. In 1902, when the company's Washoe smelter, "by far the largest source of tailings in the upper Clark Fork Basin," not to mention the largest such smelter in the world at the time, began operating, the quantity of tailings more than doubled. The tailing ponds, ditches, and dams built by the Anaconda Company to store an increasingly unmanageable amount of waste material "did little to keep tailings from washing away during spring run-off or other periods of high flow," according to Quivik. The 1908 flood deposited decades of waste behind Milltown Dam, whereas seasonal flows washed the waste downstream in what had become consistent yearly doses of poison.[43]

~ ~ ~

Although human health prompted worries about water at Milltown and along the Clark Fork, fish came to be the centerpiece of many new studies.

In May 1908, a record flood swept down the Clark Fork. Raging water demolished nearly every bridge and roadway in the floodplain from Butte to Missoula. When the river crested on June 6, nearly thirteen feet of water flowed over the dam's spillway. *Courtesy N. A. Forsyth, photographer, Montana Historical Society Research Center, Photograph Archives, Helena, Montana, ST 001.419, "Missoula Power House and Dam, June 6, 1908."*

As biologists saw themselves "beginning in the attempt to rescue the pollution-plagued river," according to ongoing newspaper stories, they focused most of their attention on "gathering information on the river's trout populations." Of particular interest were "rainbow, cutthroat, brown and Dolly Varden trout—species prized by sportsmen." Working at night with electroshock devices, biologists stunned these sport fish as they moved into shallow water to feed. Stunned trout helped the researchers "determine

Water measuring five times the average spring runoff bore millions of tons of heavy metal–laden mining wastes downriver from Butte and Anaconda. What was not deposited on the banks or swept farther downstream settled in the reservoir behind Milltown Dam, which withstood floodwaters of roughly forty-eight thousand cubic feet per second and did not fail, much to the relief of Missoula and valley citizens. *Courtesy N. A. Forsyth, photographer, Montana Historical Society Research Center, Photograph Archives, Helena, Montana, ST 001.416, "Mad Waters of the Missoula at Bonner."*

the number of fish in the river, their growth rates, species mix, age and other characteristics," such as how much they were suffering from heavy metals and other sorts of pollution.[44]

Environmental groups and the general public also saw fish health as the foremost indicator of river health. Because people shifted their concern to fish, worry over the Milltown Superfund site literally moved up and down the Clark Fork from the dam and reservoir. The issue of fish would eventually become a crucial turning point in the fate of the dam and the Superfund site.[45]

A year after repairing leaks in the dam and announcing that it was questioning the structure's fate, the MPC surprised river watchers with its vision. In mid-December 1984, the company "disclosed an $11.45 million plan to upgrade its badly deteriorated Milltown Dam on the Clark Fork River near Missoula." The company submitted a request to the Public Service Commission, which regulated energy rates in the state, to increase its electricity rates as a means of paying for the repairs. That is, the company wanted public electricity consumers to pay for a refurbished dam. In its preliminary report to the Federal Energy Regulatory Commission (FERC), the MPC claimed that the proposed project was the only way to prevent the dam from collapsing. The repairs would include "replacing the existing powerhouse and turbines," thus "increasing the dam's generating capacity to 4.3 megawatts from the current capacity of 3.4 megawatts," as well as replacing much of the dam's wood structure with concrete. The company used the growing perception of the dam as a threat to river-wide health, a proverbial toxic time bomb, to scare the public into funding its repair work. Even though Robert Periman, the MPC's hydroengineer, admitted that the dam "is nearing the end of its useful life" and that "it eventually could fail, could wash out," he rejected the idea of dismantling the dam at a slightly higher cost, which the company could not easily pass on to consumers.[46]

Others following the fate of the river seemed to agree. The regional fisheries manager for Montana's FWP thought that "upgrading the dam appears to be the most environmentally sound course of action because it would disturb the silt the least." Leaving the toxic mess in place appeared safer than trying to remove it at the time.[47] The local newspaper cheered the plan as a short-term risk with long-term safety gains. An editorial on the MPC plan saw the risks as "worth taking if, after reconstruction, the dam will not have to be lowered so often for repairs."[48]

Even with initial support from state agencies and the public, the MPC had to draft a comprehensive plan for FERC approval, which took nearly six months. When the company submitted its plan to FERC and released the two-and-a-half-inch-thick document to the public in June 1985, it reflected both widespread concern about the dam's toxic burden and the

MPC's financial dilemma. The reasoning behind the dam upgrades was to "prevent the release of heavy-metal-contaminated silt in the reservoir behind the dam." The means to pay for the $11.5 million proposal were to "raise Montana Power's electricity rates by about $2 million annually." To mend the dam, the MPC proposed building a concrete wall on the upstream side of the dam and replacing the timber cribs with a concrete shell, all meant to stop the recurrent water damage the dam had suffered. Mechanical gates would replace wooden flashboards, so that lowering and raising the reservoir level would disturb the toxic sediment less. Along with rate hikes, the company also wanted to make the dam more productive by installing new turbines that would generate roughly 25 percent more power.[49]

Not everyone embraced the MPC's plan. In Missoula, the recently formed Clark Fork Coalition (CFC) held public meetings to inform local citizens and to gauge their opinions. In fact, the CFC, which would become the foremost environmental group concerned with the dam's future, arose in response to the MPC's proposal to rebuild the Milltown Dam. Or more specifically, the CFC formed over concern about the "high levels of heavy metals" that made the dam and reservoir a federal Superfund site. As a coalition spokesperson predicted, "the Milltown Dam project will become one of the major issues affecting the river during the next year."[50]

By midsummer, with the MPC plan still in the hands of FERC, residents launched an effort to begin caring for the river on a community level. As individuals and communities were doing all over the country, Clark Fork residents created their own opportunities to remedy environmental damage. The same people who supported Superfund also supported more expedient and tangible ways of cleaning up the places they lived. The "Greater Riverfront Project," the first effort of a recently initiated forum called "Good Work Missoula," organized roughly fifty volunteers one July morning along the banks of the Clark Fork in downtown Missoula. The volunteers pulled noxious weeds, "hauled ashore the rusted pipes and cables lying halfway in the water," and generally "tugged, trimmed, picked up and carried the junk" that marred the river and marked it as a historical dumping ground. One newspaper account celebrated the diversity and cohesion of the volunteers by noting that they "were a fair reflection of the community at large. There were children and retired people, mill workers and shopkeepers, a doctor here, a lawyer over there; there were unemployed people; there were a few who were a little crosswise to the law at the moment, and there was a judge. It was kind of hard to tell one from the other; they were all sweaty."[51] After wading in the water and helping fellow community members retrieve such items as an old toilet bowl, iron rails, and a set of rusty bedsprings, event coordinator and future mayor Daniel Kemmis recognized that "there's no end to it." The Clark Fork was bringing people together but needed far more help than the volunteers could offer in a single day.[52]

Locals did their best, while some began courting more federal support. With Superfund sites already located along the Clark Fork's headwaters on Silver Bow Creek near Butte, along Warm Springs Creek where it flowed past the old Anaconda smelter, and of course in Milltown, the Clark Fork Coalition was in the midst of trying to get the entire upper river deemed a continuous Superfund complex. In June, the citizens' group pressed the head of the EPA's Superfund program, William N. Hedeman, Jr., about such an expansion. "He said he would in fact support that," said CFC spokesperson Kathy Hadley. She called the possibility "a real victory for the coalition." But this "victory" hinged in part on the ongoing studies of the river.

Rather than duplicate local, university, and state study efforts, the EPA would launch an expanded Clark Fork Superfund study only if the coordinated local study showed widespread toxic contamination. This federal response came as a direct result of the CFC's initiative, as well as what one EPA official called the complaint of many citizens "that heavy-metals contamination in the river between Deer Lodge and Milltown was being neglected." If those complaints proved valid through extant studies, the EPA appeared ready to connect the dots between Butte, Anaconda, and Milltown, making the whole upper Clark Fork a Superfund site.[53] Within a month that likelihood increased when U.S. senator Max Baucus of Montana announced that a nominee to be the new assistant chief of the EPA supported the river-long Superfund designation as part of the overall expansion of Superfund itself.[54]

~ ~ ~

The Superfund law was set to expire on October 1, 1985, but as a *Missoulian* editorial pointed out, both houses of Congress were "considering bills to extend and enlarge the program." Superfund renewal was national news. As early as February 1985, a *New York Times* article revealed that Lee M. Thomas, the new head of the EPA, supported the Reagan administration's five-year extension of the Superfund law, which would have limited its use to "hazardous waste problems and emergency spills," while eliminating its application to "mining and pesticide residues or contaminated drinking water." In line with his conservative agenda, Reagan also wished to pass the financial burden of waste cleanup programs to states. Congress, the public, environmental organizations, states, and even many corporations assailed this attempt to reduce the scope of Superfund. Accordingly, Congress passed a temporary extension of the law, while various bills to expand it bounced between legislative houses.[55]

In a political climate that forecast federal expenditure cuts, industry deregulation, and weaker environmental laws, optimism about expanding Superfund looked like a ray of hope for environmentalists, especially those along the Clark Fork. Whereas the original Superfund law (CERCLA) had established a $1.6 billion trust fund over five years for "cleaning up abandoned or uncontrolled hazardous waste sites,"[56] the new proposals in Congress would reestablish the fund for another five years with an infusion of between $7.5 and $10 billion. New versions of the law shifted more financial burden to industries besides the chemical industry, which had provided nearly 90 percent of the original fund through taxation. Proposals to expand the tax base for Superfund to more industries, including manufacturers, thus lessening the burden on those corporations singled out by the original legislation, drew support from industry. The *New York Times* quoted a spokesperson for the Chemical Manufacturers Association as saying: "A larger Superfund demands as a matter of economics and equity to be broadened." Even with some corporate support, however, the threat of a Reagan veto loomed.[57]

By the beginning of 1986, the EPA began warning Congress, the president, and the public that it would have to shut down its toxic waste cleanup within months if no new bill passed soon. The agency warned of the loss of "1500 employees working on the toxic waste program, or more than 10 percent of the agency's staff." As importantly, any further delay would stall or scuttle existing cleanup work, including studies and planning, such as was happening up and down the Clark Fork. As one EPA agency put it, "in many places we will have to put up a chain link fence and walk away." A possible halt to one of the nation's most recognizable environmental laws contrasted with the bipartisan support of its renewal.[58]

Even the political infighting between party rivals in Congress amounted to how to pay for expanding Superfund, rather than whether or not to do so. As a *Times* editorial phrased it, Superfund debate in Congress was "locked in a Lovefest," wherein "the House and Senate are competing to be the more generous benefactor." By September 1986, after a series of temporary extensions of Superfund, EPA head Lee Thomas had shifted his support away from the Reagan reductions and toward congressional expansion of Superfund. He redoubled his warning about a virtual shutdown of his agency if that expansion failed to happen soon. At the same time, both houses of Congress had agreed on nearly every detail of a new bill, save the money raising. The House favored simply increasing the tax rates on oil and chemical companies from the original law, whereas the Senate favored a broader industrial tax base. An opposing argument in Congress held that only corporations directly linked to designated toxic waste sites should pay. Either way, it seemed that an expansion of Superfund was in the offing.[59]

Superfund had not always been so secure in its short life. Just a few years earlier, as leaks in the Milltown Dam spurred conversation about its future, the future of the EPA and one of its signature programs—Superfund—came under fire because of a different sort of leak. Working at a new job as the director of environmental health for the Missoula City-County Health Department, Elaine Bild helped Milltown qualify for Superfund status by making sure all the proper paperwork made it into the proper EPA mailboxes. "I knew how the law worked," she recalled. She knew that such human agency and individual effort mattered in a bureaucratic process that included reams of official documents because she had entered Missoula government as an escape from her position as a project manager for EPA's Superfund program. Bild had left Superfund and Washington, D.C., because of the corruption the agency suffered during the first years of the Reagan administration.[60]

As Bild remembered, "the election year of 1980 brought dramatic changes for the worse." After Reagan appointed Anne Gorsuch as his new EPA director, corruption and disorganization within the agency brought it under national scrutiny. The policies and actions that Gorsuch, whom EPA employees called "Queen Anne," supported "were viewed as a reflection of Reagan's own policies." According to Bild and subsequent congressional investigations, those policies and actions included withholding agency studies and information about pollution and environmental problems, awarding government contracts for environmental work to private companies based on "who knew whom," ignoring EPA scientists and reports in favor of industry consultants, and being generally disorganized.

Bild remembered that EPA employees fed up with the situation "were leaking so much information to Congress . . . that [the] chairman of the House Energy and Commerce Committee placed cardboard boxes outside his office where the flood of documents could be deposited." "Queen Anne" Gorsuch resigned in 1983 after serving for two years as the EPA's first female administrator. While Reagan was unable to weaken the EPA or its mandates such as Superfund through Gorsuch's leadership, her years at the agency gained it disfavor in the public eye, as well as with Congress.[61]

With Gorsuch gone, and with the controversy lessened surrounding her tenure and its marring of the Reagan administration's environmental record, Superfund began to carry out its original mandate. The EPA included Milltown on its first list of 406 Superfund sites in 1983. The number of sites on the list had nearly doubled by 1985, as congressional debate about the future of Superfund focused on expansion and greater funding.[62]

During the summer of 1985, Milltown made its first mark on Superfund legislation. Long before the debate over Superfund funding was finished, U.S. senator Max Baucus from Montana, who was on the Senate committee charged with reauthorizing Superfund, tried to amend how the program worked. To the dismay of Montana environmental groups familiar with the state's mining past and the environmental damage it wreaked, Baucus attempted "to exempt mining sites from the Superfund cleanup law," according to news reports.[63] The senator defended his move by professing, "I am concerned that the current hazard ranking system may not accurately assess the relative degree of risk to human health and the environment posed by sites and facilities subject to review." Baucus claimed that his amendment would help classify types of hazardous wastes. Opponents, such as the Montana Environmental Information Center (MEIC), charged that the amendment would downgrade mining wastes in the Superfund ranking system because those wastes tended to include a variety of chemicals and elements and were often dispersed over large areas or throughout waterways. As a spokesperson for MEIC claimed, the EPA "would much rather deal with tidy little barrels of hazardous waste that they can pick up and bury." Testifying to the Senate, Baucus maintained that abandoned mining sites "could be listed on the NPL if they present a genuine and substantial risk, but certain safeguards would be put in place . . . and that higher priority sites are listed first." MEIC's past experience at Milltown and along the upper Clark Fork, as well as its concern for plenty of other contemporary and abandoned mining sites in the state, informed its opposition to the Baucus amendment.[64]

Other western states that shared Montana's mining history opposed Baucus's attempts to change the way the Superfund program classified wastes. New Mexico governor Toney Anaya condemned the purpose of the change, which he summarized as making it "difficult, if not impossible, for mining sites to qualify for cleanup using Superfund monies." Similarly, Colorado governor Richard Lamm said "no class of industrial processes should be immune from the safeguards embodied by the Superfund legislation." Both governors wrote to their Montana colleague, Ted Schwinden, hoping he would persuade Baucus to rescind his amendment.[65] Just before year's end, Schwinden too suggested that Baucus abandon the amendment. In the final product, Congress struck language singling out mining sites in favor of more general wording that defined wastes more clearly and emphasized the "quantity, toxicity, and concentrations of hazardous constituents," rather than quantity alone. Those familiar with Milltown, including Schwinden, helped retain a broad Superfund mandate by opposing Baucus's efforts to curtail the legislation.[66]

As the debate over Superfund progressed, Baucus moved with the tide of expanding Superfund, including a broadened tax base. As a key member of the Senate Environment and Public Works Committee, Baucus summarized his reasoning for a broader tax base to fund a Superfund renewal bill in a special *Congressional Digest* report on "The Superfund Controversy" by writing, "Everyone contributed to the problem, and therefore, everyone should contribute to cleaning up the problem." That included mining. He went on to confirm that Superfund was an accounting for America's industrial past, which justified making current industry pay, quipping that "past waste management practices allowed us to manufacture goods for less than the true cost of production." Superfund renewal would allow collection of those historic debts, of which Montana had many.[67]

~ ~ ~

By the summer of 1985, the Montana governor's Clark Fork River Basin Project, charged with participating in the ongoing river studies, was making predictions about the waterway's fate. Recognizing that the Clark Fork, like most western watersheds, had hosted "more than a century of mining, timber, agricultural and municipal discharge," Ken Knudson, the head of the study, declared that the river was incapable of cleaning itself up in any appreciable time frame. Therefore, river cleanup was the responsibility of its longtime users. The first hurdle for the Clark Fork, according to Knudson, was to get people to see the river as something other than an industrial sacrifice zone. "For so long not only did people take their streams for granted, but they considered the Clark Fork Basin a lost cause because of the complexity and magnitude of the problem," said Knudson. So, along with studying the specifics of how polluted the river was, the first step toward the river's future was creating a new vision of it. With people flocking to the riverbanks to help haul trash out of its waters, and with the coordination of local, state, and federal studies, both of those initial steps seemed to be taking place. In July 1985 President Reagan's nominee to be the new assistant administrator overseeing Superfund announced that he supported providing Superfund money to study the entire Clark Fork basin, which could lead to more Superfund designations along the river.[68]

As a river-wide focus coalesced along the Clark Fork, Milltown once again took center stage. The late February flood of 1986 that carried a heavy burden of ice down the Blackfoot and Clark Fork and broke wooden gates along the top of the Milltown Dam returned the discussion of the river's future to its central Superfund site. As flooding subsided, the MPC lowered the Milltown reservoir nearly seven feet, shut off the dam's turbines, and hosted officials from FERC to begin assessing damage to the dam. The flood had burst fifty of the sixty-seven wooden gates and smashed a "living-room-sized" hole "in the apron on the downstream side of the dam" where water crested the broken gates and pounded down its face. An MPC spokesperson warned: "We don't think it will wash out immediately. . . . But left unattended, something could happen in the future." The employee speculated that the flood damage might prompt FERC to hasten its approval of the company's $11.5 million proposal to rebuild the dam. Whatever the status of its long-term reconstruction plans, the MPC proposed immediately repairing the downstream apron, which covered and protected the dam's rock and timber crib foundation. A ruptured dam could spill the estimated seven million cubic yards of toxic sediment into the entire fishery below Milltown, including the more populous Missoula from where the CFC and most of the environmental concern stemmed.[69]

While the MPC considered its options to restore the dam, the flood-damaged structure amplified the conversation about the river's future. A month after the flood and with another, more seasonal spring surge of high water expected soon, Governor Schwinden attended the Clark Fork Coalition's annual meeting in Missoula to address the river's future.

Acknowledging that some industrial use of the state's resources would continue, the governor also emphasized that "there is an overwhelming desire among Montanans to preserve a place worth living in." Crediting the CFC with arousing most of the concern surrounding the Milltown Dam and the upper river, he claimed that cleaning them up was "among the most exciting and worthwhile possibilities of my governorship." He emphasized the need for ongoing research to quantify the river's problems. The newspaper article devoted to the gathering made the first public reference to "restoring" the river. Schwinden promised the state's continued support in that effort, while maintaining that "a spirit of shared responsibility is essential if we are to successfully tackle the river's very complex problems and the issues that are related to the recovery of the river system." Looking both backward and forward at the cleanup project, he finished with the platitude: "As we approach our Centennial, we can feel proud that in Montana's second century, the Clark Fork will gradually be restored to what a river should be: A source of life and, just as important, a source of inspiration to those whose lives it touches." The head of the governor's river study reiterated the forward-looking nature of restoration by urging members of the CFC to lay plans for the river's future.[70]

The most immediate of those plans centered on the MPC. A month after the February flood, the company warned that "potential loss of life or property" should induce FERC to "accelerate" the review process for its $11.5 million renovation plan. While state officials and the Clark Fork Coalition shared the worry about the dam's failure, they were less inclined to accept a hasty decision on its fate. The post-flood turning point for the dam's future came to a head over the company's desire for quick approval of its plan versus state and local opposition to such expedience. The governor's Clark Fork River Basin Project and the CFC both warned FERC that they would "formally intervene . . . in an effort to force a full public review of the reconstruction project." CFC spokesperson Peter Nielsen took the long view of the circumstance, saying, "We're talking about what we'll be living with for 50 years. . . . Let's make the right decision."[71]

While spring runoff never swelled enough to damage to the dam further, the issue of how to repair it remained. In what seemed like a break from local environmental groups, the state Department of Health and Environmental Sciences (DHES) proposed suspending state water-quality standards to allow the MPC to repair the dam while disregarding the amount of sediment washed downstream in the process. The DHES move was meant to help the MPC gain approval for dam renovations by FERC, which required that any such work would not violate state water-quality laws. Knowing that work on the dam would likely do so, the DHES offered to simply curtail such laws. State fisheries biologists, the CFC, and MEIC pounced on such a suspension of state environmental regulations and threatened that a full environmental impact statement should be required for any dam rebuilding project.[72]

By the first of May, FERC seemed to ameliorate the debate. Acknowledging state and local concern, the agency announced that it might grant the MPC permission to repair the dam's damaged spillway during the coming summer but would hold its $11.5 million proposal to rebuild the entire dam until federal regulators had sufficient time to study the problem of potential toxic releases more thoroughly. This new two-part process would also allow the completion of the Superfund study before any decision was made. The CFC's Peter Nielsen called FERC's phased review "good news." The local newspaper, the state, and even the MPC echoed Nielsen's evaluation.

FERC approved the spillway repairs within a month, leaving in place Montana's water-quality laws and requiring the MPC to monitor water quality throughout its work.[73] The approval required the MPC to take all possible precautions against spilling toxic sediment downstream. More importantly, it left in question the possibility of rebuilding the entire dam.

Throughout the dam's history, leaks and floods eroded, dinged, and damaged it. Following the discovery of arsenic in Milltown wells, the threat the river posed to the aging dam expanded the concern about the toxic sediment it impounded. The flood of 1986 brought to a head the issue of the Milltown Superfund site as a river-wide worry. The expanded space of concern coincided with an expansion of people involved in the issue, including new local environmental groups and the state as well as the dam's corporate owners and federal agencies. While these groups and individuals agreed to immediate repairs of the damaged dam, they also began serious debate over the structure's long-term future.

CHAPTER 3

Placing the Blame

Liability at Milltown and the New Superfund

Before the flood of 1986, the contaminated reservoir garnered most of the attention at Milltown. The flood shifted worry to the poor condition of the Milltown Dam and what its failure would unleash on the river. At the same time, the Superfund process heightened focus on the impoundment. After the initial problems of addressing the contaminated drinking water and identifying the source of toxic wastes were solved, Superfund protocol called for the identification of potentially responsible parties (PRPs) and the drafting of a feasibility study (FS). The heart of the FS was a list of possible long-term solutions, or "remedial actions," for the problem, which came to include dealing with the dam. As a faulty stopper in the river, the dam was suddenly as big a problem as the water it held.

When the Clark Fork's flow dwindled to its yearly low in mid-August of 1986, contract workers for the Montana Power Company began repairing the Milltown Dam's flood-damaged spillway. Because of the reservoir's status as a Superfund site, the Federal Energy Regulatory Commission (FERC) granted the MPC quick approval for structural repairs to the dam. Since there was an "emergency," FERC, as well as Montana's state environmental agencies and local environmental groups, favored patching the dam as soon as possible, even at the risk of releasing some toxic sediment from the reservoir into the river. Failing to act posed the bigger risk of a complete dam failure.

While the MPC contractors working on the dam expected to finish repair work within a year, the company also hired water-quality monitors to analyze how construction impacted the river. The MPC's permit required both the time frame and the monitoring because, as FERC wrote in its assessment, "the potential for discharging large amounts of toxic-laden sediment downstream of the spillway during construction . . . is great." The local newspaper warned that the dam "was an environmental disaster waiting to be unleashed." Those worries took shape around the MPC's reconstruction of its dam. As that work progressed, many of those same observers, particularly members of the Clark Fork Coalition (CFC), a local environmental watchdog group, praised the MPC and its contractors for avoiding disaster. Throughout the $2.3 million project, water-quality monitors detected only "minor increases in metals" well below permitted limits.[1]

Success in remodeling the dam, or what the MPC imagined as "Phase I," prompted the company to speculate on when it might begin "Phase II," the centerpiece of which was an upgraded powerhouse. But that speculation soon ran into reality.

In particular, the dam owners faced mounting opposition from environmentalists. As a graduate student in the University of Montana's Environmental Studies Department, Peter Nielsen got interested in a lawsuit by the National Wildlife Federation and Trout Unlimited against the MPC's plan to draw down the Milltown Reservoir at a pace and to a level that the plaintiffs claimed would harm fish and aquatic life. Before the discovery of arsenic and other heavy metals in those sediments, Nielsen followed the lawsuit and joined protests supporting it with a focus on the quantity of sediment that rapid drawdowns sent downriver. An injunction stopped the MPC from proceeding with its plans; the experience also changed Nielsen's career path. He stuck around Missoula after graduate school to be a whitewater rafting guide. Then in 1985 he helped organize and guide a whitewater outing to raise money for the people organizing the CFC.

By January 1986, Nielsen was the CFC's first permanent executive director. The group envisioned itself as an environmental organization working for the good of its namesake watershed. As such, the coalition was on the leading edge of a wave of new, locally grown conservation groups across the United States focusing on whole river systems. Like others in the burgeoning watershed movement, Nielsen emphasized science. He and fellow coalition members also courted business. In a letter to the Army Corps of Engineers about the MPC's Phase II plans, Nielsen described his group as "an alliance of 53 businesses and organizations." Business support was part of the coalition's goal of being a community-wide organization, rather than an issue-based grassroots group. Businesses joined because it gave them proenvironmental credibility and a small tax deduction for their support, and because many local business owners were outdoor enthusiasts, much like Nielsen. In 1988 an *EPA Journal* article on the relationship between commercial interests and public environmental opinion stated that "for business to maintain its profitability, influence, and freedom, it must be sensitive to the concerns of the public—not just in terms of the price and quality of the goods it produces but also in terms of public approval of its social and political influence." Support for groups like the Clark Fork Coalition was a visible, local avenue by which a business could reveal "its social and political influence."[2]

One of the CFC's largest business supporters was the Montana Power Company. Yet, as the Milltown Superfund site became the group's most prominent concern, it strained the coalition's relationship with the dam's owners. Moreover, it was around that strain that the CFC thrust itself into the public eye.

As the dam gained attention at Milltown, so did the issue of who was responsible for cleaning up the toxic mess. EPA authority and local advocacy combined to push the MPC toward becoming part of the long-term solution at Milltown and put the emphasis on what to do with the dam, rather than just on how to clean up the contaminated sediments behind it. Where the MPC saw fixing the dam's spillway as the first of two steps in improving its asset, the Superfund law deemed it an emergency response, similar to providing Milltown residents with a new, safe drinking water system. While the MPC wanted to move on to further improvements, the EPA, state agencies, and environmentalists began pursuing the next phase of the Superfund process—establishing liability and investigating permanent remedies.

At the same time that the EPA was enforcing its authority over corporate responsibility for cleaning up the environment, it watched the fate of one of its signature programs—Superfund—go from hanging in the balance between congressional support and executive veto to expanding far beyond its original mandate and budget. Although this was happening far from the impounded waters of the Clark Fork, the situation at the Milltown Dam and Reservoir continued to alter how the public participated in the Superfund process. All the while, the Superfund process began reaping benefits for the contaminated river, long before any official remedial actions happened.

In addition, Superfund's trajectory at Milltown and at the national level supports scholarship showing that despite the antienvironmental rhetoric of the Reagan era, it was during the 1980s that most of the significant federal environmental laws passed in the 1960s and '70s underwent their first significant implementation. In the case of Superfund, Congress strengthened the law. On the other hand, environmental historians have tended to see great polarity around environmental issues and laws during this decade, whereas a close look at a place like Milltown provides an example of how federal and state agencies, environmental groups, the general public, and, at times, major corporations embarked on a path of compromise. Furthermore, Superfund law encouraged, even mandated, such compromise.[3]

These results at Milltown confirm some arguments about national environmental history in the 1980s. Historians such as Samuel P. Hays have portrayed this decade as a time when support for the nation's ten largest environmental groups boomed and they moved to Washington, D.C., and became major lobbying organizations. As Hays and others have argued, those changes resulted from widespread, long-term national support for environmental issues inflamed by Reagan-era antienvironmentalism. When the executive branch did its best to cut environmentalists out of national politics, their popular support increased. So, too, a window of opportunity opened for local grassroots groups. But it was at that local level that Hays's assertion that "environmentalists were rejected as legitimate participants in the give-and-take of public affairs" makes less sense. When participation in public affairs is reduced to large environmental groups and top administrators of the EPA, those groups may have experienced rejection. Viewing the happenings at a place like Milltown leads to a different conclusion. Hays argued that environmentalists found "the administrative apparatus of the state closed to their influence." As this western Montana Superfund story makes clear, the Clark Fork Coalition worked closely with the EPA. And the agency, the authority of which resided at the regional, if not local, level, courted the participation of local environmentalists and the public. That is, Milltown illustrates how local environmental groups may well have filled gaps that larger organizations deemphasized but did so largely in concert with federal and state agencies. Superfund abetted environmentalists' participation in public affairs. And groups like the CFC did so because of their long-term commitment to community needs, including a host of real, local environmental issues, not because they were reacting to national politics or a single environmental problem.[4]

Finally, the idea that the Reagan administration provoked a transfer of power from the federal government to the states, especially in environmental regulation, fails to reflect how Superfund was working. When and if the state gained authority, it was alongside a strengthened and better-funded federal government (the EPA in this case). The EPA drew on and encouraged state authority. There wasn't a finite pie of power over which the state and the federal government fought for greater portions. Rather, all involved parties, from the federal government to individual citizens, gained new authority or expanded roles in cleaning up the environment through a (mostly) cooperative process.

One thing that historians have seen clearly is the motivation people brought to environmental issues. As Hays observed, "the expression of environmental aspiration was infused with a sense of place. People thought of environmental quality in terms of where they lived or worked or engaged in recreation." That was undoubtedly true at Milltown, although the parameters of the place in question were shifting.[5]

~ ~ ~

As early as the initial designation of the reservoir as a Superfund site in 1983, the EPA foresaw having to deal with the dam. At the time, the agency already viewed the dam as a "source-control," a tool for dealing with the toxins it trapped. The MPC's maintenance and operation of its dam mattered. In the fall of 1984, the EPA stamped its authority on the situation in a letter to the MPC, stating that the company must communicate "any changes in the dam structure or operation" to the EPA and to the state's Department of Health and Environmental Sciences. More pointedly, the EPA threatened the MPC with clauses in the Superfund law that could result in the company becoming a responsible party for cleaning up the site if its actions released toxic sediments from the dam. EPA Region 8 director John Wardell wielded a financial stick. If the EPA designated the dam owners as a potentially responsible party (PRP), the company would have to pay an agency-determined portion of future cleanup work. If the company refused its responsibility and the EPA had to foot the initial bill, the federal government could sue the company for up to three times the cost the EPA spent on cleanup efforts.[6]

The power company began informing the EPA of its plans. The MPC's preflood engineering evaluation for rehabbing the dam included renovating the powerhouse to increase its power production. The company planned to make repairs and upgrades at the same time. It also prioritized minimizing long- and short-term environmental impacts, such as allowing toxic sediment to wash downstream. As part of that effort, the MPC considered excavating contaminated sediment exposed by lowering and rerouting the reservoir during construction.[7]

After the flood of 1986 damaged the dam, plans changed. By the end of March, the EPA and all of Montana's environmental agencies—the Department of Health and Environmental Sciences (DHES); Department of Natural Resources and Conservation (DNRC); Fish, Wildlife and Parks (FWP); and Clark Fork Coalition (CFC)—petitioned FERC to intervene in the dam reconstruction licensing process. The EPA paved the way for this intervention using Superfund law. The law empowered the agency to enter any legal proceedings that might involve the release of hazardous wastes from a Superfund site.

Each agency had specific reasons for intervening. In general, the state agencies wanted to safeguard human health (especially the downstream Missoula aquifer that was the drinking supply for much of the city) and the river's natural resources against releases of heavy metals. The EPA added to those concerns its future decisions regarding the Superfund site. The CFC insisted on "adequate public review and intergovernmental cooperation," an independent, on-site monitor to assess downstream water quality during construction, and the implementation of an MPC-funded fish mitigation program. All the interveners favored FERC dividing the company's proposal into two phases: fixing the ruptured and near-failing spillway, and upgrading the powerhouse.[8]

In a late spring news release, the DHES announced that it, along with the EPA, had approved a FERC "emergency" order to allow the MPC to repair the dam's spillway. But the state and federal agencies pushed FERC to disallow the power company's plans to upgrade the powerhouse and generating capacity at the same time. Phase I could and should proceed immediately; Phase II would have to wait. The justification was that the EPA's study of how to deal with the contaminated reservoir (the feasibility study) was due out within a year, and one of the possible Superfund alternatives was complete dam removal. If the EPA included dismantling or altering the dam as part of the solution to remediating the reservoir, the MPC's improvements would be a waste. The EPA and state environmental agency were unsure about whether the power company could sue for lost investment. What both agencies knew for sure was that the more repair and upgrade work the company undertook on the dam, the more likely its efforts would result in releases of contaminated sediments into the river.[9]

That wasn't the only problem the EPA faced at the time. In fact, much bigger changes loomed over the agency.

~ ~ ~

The original Superfund legislation, passed by Congress in 1980, expired in 1985. A temporary measure kept the program funded through midsummer 1986. Just as the EPA began dangling Superfund laws over the MPC, the agency was in jeopardy of having the program's funding and authority run out. Along with the hundreds of Superfund sites around the country, Milltown faced abandonment by the federal government.

By October 1986, a Superfund reauthorization bill was in front of President Ronald Reagan, and his advisers were recommending he veto it. Upon entering office, Reagan and his administration had tried and failed to undermine the original Superfund mandate. Now the president faced a congressional bill that increased the program's funding fivefold to $8.5 billion over a new five-year period, which was down considerably from some of the earlier iterations of the bill. Many in Congress had acknowledged studies by the EPA that showed it would take $22.7 billion to clean the country's 786 top-priority Superfund sites, to say nothing of the nearly 20,000 places the agency forecast as needing designation. The Congressional Office of Technology Assessment's estimate for the same work topped out at $100 billion. The bill that quintupled the Superfund budget would come from taxing a much broader base of industries than the original legislation had. Both heavier taxes on corporations and public spending on the environment defied Reagan's stated economic principles. Yet he faced strong support for the bill in Congress (even conservatives recognized how conservative the new fund was) and among EPA administrators, national environmental groups, and the Chemical Manufacturers Association and other industries (which realized how much steeper the bill could be). Everyone except the executive saw Superfund as a way to deal with environmental disasters more efficiently and cost effectively than leaving the task to government agencies or, conversely, private business alone.

On October 17, 1986, Reagan signed the Superfund Amendments and Reauthorization Act, or SARA. For all the antienvironmental rhetoric and efforts that characterized Reagan's first few years in office, strengthening one of the nation's seminal environmental laws made seemingly little impression on the president at the time. In 1983, when EPA administrator

Anne Gorsuch faced corruption charges in Congress, Reagan defended her in his diary against "a lynch mob" media that "thinks it smells blood." He dubbed congressional Democrats investigating EPA misdoings as "some zany on the hill." A year later, he privately accused environmentalists of having "declared war" on him at the beginning of his first term and of being "arrogant and unreasonable" in general. Yet Reagan recalled nothing about signing SARA in his diary, instead commenting that he had "forgotten how really good" the 1942 film *Kings Row*, which helped him rise to acting fame, was. Although Reagan took little note of the new law, some of its amendments began to play out at Milltown.[10]

Even with its budget increase, the EPA continued pursuing its "polluter pays" strategy for funding cleanups. At Milltown, the agency began designating potentially responsible parties (PRPs). The law defined a PRP as any individual or corporation that might be liable for cleaning up a Superfund site. At Milltown the agency singled out the Atlantic Richfield Company (ARCO), Champion International Corporation, and Montana Power Company. Each PRP had a different link to liability.

The EPA named the Atlantic Richfield Company as PRP number one because of its connection to the upper Clark Fork's industrial history. At the headwaters of that river Butte grew from a cluster of gold-mining shacks in the 1870s to one of the West's largest industrial mining centers, producing 43 percent of the nation's copper by the time Montana became a state in 1889. The vertically and horizontally integrated mining industry dominating Butte included huge ore smelting works in nearby Anaconda, named for the company that between 1900 and 1910 had consolidated ownership of most of the Butte mining, smelting, and refining works. According to the EPA, "as a result of one or more mergers, restructurings, transfers of assets, continuation of business activities, or other corporate actions, ARCO is the successor-in-interest to, and has assumed the liabilities incurred by Anaconda." Companies or individuals inherited liability.[11]

When ARCO purchased Anaconda assets in 1977, including the smelting properties in the Deer Lodge Valley, the Clark Fork floodplain from Silver Bow Creek to Milltown was defined by stretches of barren stream bank, called "slickens," as much as it was by native willows and cottonwoods. The Milltown Reservoir alone contained nearly two thousand tons of arsenic and an equal weight in lead, as well as tens of thousands of tons of slightly less poisonous metals such as manganese, zinc, and copper that had escaped extraction. In 1977 there was no Superfund law and people knew few of the details about the extent of contamination in the Clark Fork, so ARCO had little reason to consider its environmental liability.

Superfund law included the transfer of liability for such messes. Companies bought liability along with their acquisition of assets. This "strict liability" allowed the EPA to designate corporations, government agencies, or individuals without "showing of fault or negligence on the part of a defendant." Accidents counted. The law was "retroactive and progressive." Toxic messes made long before or after passage of Superfund fell under its purview. Time did not matter. And liability was "joint and several," meaning that any defendant responsible for some portion of the waste problem could be responsible for the entire problem, unless it proved that the pollution was "divisible." The new Superfund law strengthened all of these clauses, making elusive responsible parties less so. The law's supporters realized that, as at Milltown, toxic pollution often hid behind a proprietary web stretching back through a century of American industrialization. Even ARCO recognized its responsibility under these laws and participated as a PRP from the outset, albeit under threat of the treble damages that CER-CLA allowed the EPA to charge if the agency had to fund cleanup and sue PRPs for reimbursement.[12]

For its part, Champion International Corporation denied any responsibility connected to the Superfund site. A company spokesperson characterized his employer as "an innocent owner of a minute portion of the property below the high water mark of the reservoir." In addition, he explained to the EPA that while Champion had "generated" none of the offending toxins, it had voluntarily helped Milltown residents acquire a new water system and had contributed to Superfund cleanup efforts elsewhere, when the company "was properly identified" as a PRP. That, according to Champion, did not include Milltown.[13]

The Montana Power Company's connection to the contamination was an exception. The company had bought the dam in 1929. It never produced or transported any of the reservoir's toxins. But, according to the Superfund Amendments and Reauthorization Act, intentions didn't matter as much as results. Beginning with the construction of the dam in 1906, the structure had caught and accumulated heavy metal–laden sediments, especially during the flood of 1908. The dam, by creating the reservoir, was responsible for the conditions that conveyed those heavy metals into the surrounding groundwater and from there into people's homes. Maintenance of the dam included flushing sediments downstream. Plus, the

MPC actually owned portions of the land on which the reservoir pooled; hence the company was liable for the sediments thereon. Yet the MPC denied its liability. Early in the Superfund process at Milltown, that denial had legs. SARA's general provisions about liability seemed to implicate the MPC, but a single anomaly let the company off the hook. In one short paragraph buried in section 118(g), SARA absolved the Milltown Dam's owners and only that dam's owners of liability for the toxins behind the structure. That exception was the work of Montana senator Max Baucus, who had helped craft the SARA bill. But it was a short-lived exception that Congress struck from Superfund law before the EPA made its final decision at Milltown.[14]

Even as the EPA designated PRPs at Milltown, the agency began more immediate forms of enforcement. In particular, the Superfund law encouraged the EPA to draw on state support. In that effort, the EPA assigned ARCO the responsibility of evaluating the state's legal power to implement what Superfund termed "institutional controls," or "legal mechanisms to regulate land use and access to prevent exposure to hazardous substances." This included laws such as zoning restrictions or water regulations, which the EPA preferred states handle. But as the federal agency confided to Montana's director of the DHES, Superfund implementation had taught the EPA "that many States and localities lack adequate authority, resources, or commitment to make institutional controls a viable component of remedial actions." Superfund encouraged the EPA to help states define, fund, and embrace their role in carrying out institutional controls. The EPA saw state aid as part of a "temporary remedy," one that the state could put in place quickly while the permanent remedies of the Superfund process played out. Increasing state involvement in the Superfund program was one part of SARA that seemed to reinforce the Reagan administration rhetoric of shifting regulatory power toward states. In truth, the situation at Milltown demonstrated that the EPA sought quick and efficient enforcement of the Superfund law by working *with* the state, rather than shifting power from the federal to the state level.[15]

State agencies, of course, had already participated in the Superfund process at Milltown and continued to do so. Most of that participation continued to concern the MPC's licensing process, first as the dam's owner petitioned to carry out Phase II rehabilitation work, and then as it applied for special extensions for operating the facility.

The company's hopes of refurbishing the dam's powerhouse and increasing its power output faltered because of the reservoir's Superfund status and its own potential responsibility. All of the state agencies involved (DHES, DNRC, and FWP) ultimately opposed the improvements, as did the CFC and the EPA. In the short term, they all worried that further work on the dam risked flushing contaminated sediments downstream. And since they did not consider Phase II an "emergency," the organizations wished to avoid that risk. In the meantime, all agreed that since the MPC was a PRP (albeit unwillingly), and since the FS, which might determine the dam's ultimate fate, was due out within a year, the dam owners should cease working on the structure. What no one suspected at the time was that the dam would become the centerpiece of cleanup efforts and that the FS would take much longer to produce than expected.[16]

The MPC tried to counter criticisms of its plan. The company emphasized how Phase II did, in fact, include structural upgrades that would make the dam safer and less likely to discharge its toxic holdings. Engineers blamed any such releases during Phase I on "natural events" rather than on construction. At one point the MPC's vice president of engineering wrote that "changes in water quality are often brought about by natural occurences [sic] and are independent of pond level," as if the tons of heavy metal piled behind the dam had no bearing on the matter, or occurred naturally. As far as the environmental impacts of these occurrences, whether natural or not, the company claimed that a century of upstream mining had damaged the river so profoundly that what happened with the dam was all but negligible. Using fish as an example, an MPC report claimed: "We do not agree with the conclusion that the presence of the dam significantly reduces the recruitment of young fish to downstream areas." The assertion that the whole river was a wreck because of historic mining, and that the dam was of no consequence, was meant to imply that the MPC had little or no responsibility under Superfund. All the while, the company seemed to be trying to diminish the magnitude of the problem by referring to the site as the "pond" rather than the much-bigger-sounding "reservoir," which was in standard use at Milltown. Again, that may have been another subtle attempt to escape PRP status.

The one portion of the Phase II plan that eventually roused the most local opposition was the MPC's choice of where to dump toxic sediments. During Phase I "emergency" repairs, the company dredged sediments and dumped them along with old, possibly contaminated, dam materials in what was supposed to be a temporary disposal site near the reservoir. That disposal site became another point of contention in the debate over the second phase. In the fall of 1987, John Wardell informed FERC that the MPC's disposal site "appears to be located within a floodplain and wetland." At a place where water contamination had earned Superfund designation, that was not good. As Montana's EPA director, Wardell laid out the hazards of dumping even more sediment there if FERC permitted further dam rehab. The site had no liner to prevent the movement of contaminants back into the surface or groundwater, nor was the site safeguarded against floodwaters, and the disposal site's placement in a floodplain or wetland exposed many of the sediment's fairly inert metals to the air (or oxygenated water), which could, according to Wardell, increase their "toxicity and mobility." Wardell reminded FERC that since this was happening in a Superfund site, the EPA had ultimate authority, especially since the issue concerned a PRP possibly spreading more contaminants. Facing identical concerns voiced by the Clark Fork Coalition, the MPC began adjusting its plans.[17]

In October 1987, representatives from all the concerned parties convened for a "Milltown Dam Rehab Meeting." In spite of all the initial opposition, FERC and the MPC seemed to make a clear case that the old dam needed further repairs for safety reasons, which did not constitute an emergency but were urgent nonetheless. Since the EPA officials admitted that a feasibility study was "no nearer being finished now than we were a year ago" and might not be complete until "mid-1989," that urgency would only increase. Everyone agreed that the Phase I disposal site was chosen for its proximity and practicality during emergency repairs and would not be suitable for further disposal. The meeting ended with the EPA warning the company that if repairs progressed and the FS eventually scuttled those repairs, there was no guarantee that the MPC would receive any compensation.[18]

The one projection that the MPC got right was its rebuttal to stalling Phase II because of the EPA's FS. The FS was suffering delays and the company found it unfair that its rehab permit should remain on hold indefinitely. And, had FERC simply put the Phase II permit on hold until finalization of the FS, the rehab work would have sat in limbo for another decade.[19]

In 1987, the EPA delayed the FS by deciding to focus on cleaning upriver sites first before making a final decision about more work at Milltown.

With Milltown's FS essentially on the slow track, the MPC pushed even harder for its proposed rehab work, making use of what the Clark Fork Coalition called in its fall newsletter, *Currents,* "the same old scare tactics about instability of the dam." The coalition was particularly upset about the MPC's desire to forgo an on-site monitor and to continue using a wetlands disposal site for sediment storage. All noncorporate parties opposed those choices, too. During a meeting that included the EPA, FERC, state agencies, and the CFC, agency and local pressure finally won out. In the winter edition of *Currents,* the CFC revealed that the MPC had acquiesced to the demands for a safer, though more costly, disposal site, one about six miles upstream and well out of the river's floodplain. In a demonstration of its power through "institutional controls," the state continued to insist that the MPC meet water-quality standards throughout the proposed project, which would likely force the company to treat river water passing through the construction site.[20]

The CFC, directed by Peter Nielsen, had been involved at Milltown since the beginning of its Superfund designation. But it was only in the late 1980s, during the debates that put the dam at the forefront of the cleanup process, that the coalition became *the* most important and publicly recognized local voice in those debates. It did so while enjoying the support of local citizens and businesses. Just as it courted diverse support, the coalition spread its attention over a variety of issues connected to its namesake watershed. Thus, it diverged from the image of traditional grassroots environmental groups.

~ ~ ~

The Clark Fork Coalition's influence was clear by the spring of 1988. In March, the *Missoulian* announced that FERC had approved dam improvements "to protect safety, but the company cannot install an additional megawatt." Power upgrades would have to wait until after the EPA reached a decision on cleanup. The MPC also faced all the issues and requirements that the CFC had argued for: an on-site monitor would be present during all construction, water-quality standards would remain in place, and a new off-site disposal area would replace the wetlands dump. The local newspaper's editorial called the decision "a remarkable result," crediting "compromise" for defusing "a heated environmental controversy." But that compromise soon began to unravel.[21]

Soon after the FERC ruling, the MPC appealed its new permit, asking for FERC to extend its operating license for the Milltown dam, set to expire in 1993, until 2015, as well as for a guarantee that it would be able to increase its power generation before the original expiration date. According to Peter Nielsen, the appeal also sought to weaken language about an around-the-clock monitor and water-quality standards during construction. As the new appeal sat, the MPC moved ahead with actual work. By midsummer it was drawing down the reservoir by a foot per day, a much slower pace than usual, as required by its permit to protect against flushing toxins downstream. The reservoir would be drained and ready for construction by July 18, according to dam engineers.[22]

One reason that the details of the dam rehab garnered so much meticulous attention locally was that concern for clean river water was in the news. As the MPC was appealing to weaken the restrictions in its permit, the city of Missoula was trying to get its water supply system designated as a "sole-source aquifer." That federal designation signified that the municipality obtained nearly all its potable water from a single groundwater system. And under the federal Safe Drinking Water Act, sole-source aquifers qualified for "an elevated level of protection." That protection included a restriction on federal dollars going to any project that might contaminate a sole-source aquifer, including money spent on Superfund sites. So, if the city's bid succeeded, it would force the EPA to enforce the strictest measures against releases of heavy metals upstream of Missoula. At the same time, the city was working to clean its own water resources. Working with the CFC, the city requested a stricter pollution control permit from the state Department of Water Quality. The new permit increased water monitoring, set stricter sewage treatment standards, and, most importantly, set Missoula on the path to follow a strong national trend of banning phosphate detergents. Nielsen's CFC helped both measures succeed.[23]

Locals concerned with their water saw the city's arterial river, the Clark Fork, looking unseasonably dirty in mid-July 1988. The cause was the MPC's rehab project. After receiving its final FERC permit on July 8, the company then had to hire a contractor and obtain state building permits, which delayed the start of construction even as the reservoir sat empty. The first phase of work was the building of a temporary dam, or cofferdam, above the permanent and problematic structure to prevent the sort of sediment scouring and flushing that was causing the river to appear dark and ominous, according to the CFC. Company officials, and eventually state water-quality monitors, assured the public that the sediment-darkened river posed no human health risk. At the same time, state fisheries biologists worried that the sediment could affect the downstream trout population by killing insects on which they fed. Whether it was bugs, fish, or people that were at risk, the CFC and the EPA began taking independent water samples below the dam.[24]

The MPC's contractors commenced the $13 million project by building the cofferdam just before the end of July, which helped clear the river's water. That did not stop the CFC from filing a formal complaint with the EPA about "a potential for continued release of hazardous substances from the Milltown reservoir." The complaint mustered evidence correlating the dam drawdown and delay with the sediment flows and requested EPA assistance with analyzing the coalition's independently gathered water and sediment samples. Arguing that it should not "be the responsibility of a citizens group to pay for lab analysis when hazardous substances have been released from an NPL site," the coalition sought to draw on the continuous and cooperative relationship it had developed with the federal agency through the Superfund process.[25]

But good relationships with the federal government did not trump regulations. The EPA had Montana's DHES respond to the CFC, explaining that since the federal agency had already empowered the state agency to sample and analyze river water from the same sites during the same times that the coalition sampled the river, the environmentalists' efforts were both redundant and unauthorized within the EPA budget and plan. However, the DHES hinted that the samples had greater value than just figuring out how much damage the one-time sediment flush did to the river. The EPA planned to include the information in its ongoing efforts to write an FS, which might include "drawdown of the reservoir and/or dam retirement" as permanent solutions for Milltown. The letter also revealed that of the three PRPs the EPA had designated that spring, ARCO was the only one that had "responded positively" and "expressed an interest in taking the lead at Milltown." That lead taking included sharing the responsibility of producing an FS.[26]

As agreeable to its role of responsible party as ARCO seemed to be, the MPC remained problematic. As the EPA expedited the lab results from samples of downstream river water, the agency also requested that the dam's owners speed up their construction schedule. Hurrying up construction, the EPA reasoned, would diminish the likelihood of subsequent sediment flushing and hence the likelihood of having to extend Superfund designation below the dam. The MPC continued to balk at the notion that substantial toxins had flowed below the dam during its initial drawdown. And a company spokesperson rebuffed the EPA request, saying that the company's work schedule was set because of its dependence on the availability of construction materials.[27]

~ ~ ~

During the summer of 1988, as the MPC wrangled with the environmental community and various agencies, the company also continued to deny its role as a responsible party. From a funding standpoint, that meant that the EPA would have to enforce a "fund lead" liability scheme. This meant that the federal agency would pay up front for any expenditure at the Superfund site and then try to extract reimbursement from PRPs later, through negotiations or adjudication. If the EPA ended up in court with a PRP, Superfund law entitled the government to charge the defendants up to three times the amount of the Superfund bill the agency incurred. That method of funding cleanup work characterized Superfund's earliest efforts, which were terribly inefficient. What was happening at Milltown represented a changing liability scheme.

After Congress passed the SARA amendments in 1986, the EPA began pursuing an "enforcement lead" strategy at Superfund sites. Even though the new legislation more than quintupled federal funding for cleaning up the nation's worst toxic messes, the EPA began using its authority to force PRPs to fund cleanup work from its first stages through its completion. Studies showed that assigning corporate responsibility early on eliminated or reduced court time, reduced the cost and time of actual cleanup, and kept Superfund in the black.

Reducing costs at cleanup sites was particularly important where mining waste was the source of contamination. Mining sites tended to cost more than twice as much to clean up as any other type of site. At Milltown, no one knew what the final bill would be in 1988, but since the EPA first named its PRPs, ARCO, in contrast to the MPC, had accepted its responsibility. By doing so, the company participated in the EPA's transition toward "enforcement lead" cleanups. Ultimately, ARCO bore the lion's share of the cleanup costs and responsibilities. But for the first few years following the flood of 1986, most of the public attention and interorganizational tensions involved the MPC. Even in the EPA's 1988 "Initial Site Evaluation," the agency scarcely mentioned ARCO, whereas it detailed the ongoing permit negotiations between the MPC and FERC.[28]

By the fall of 1988, the Montana Power Company faced further setbacks in its efforts to relicense its dam and in its public perception. Initially, the dam owners and the CFC made attempts to mend their relationship over the issue of fish. The environmental group helped broker a deal between the company and the state in which the MPC would contribute $50,000 to study the fishery downstream of its dam in an effort "to boost the river's fish population." In exchange, the coalition and the state agreed to support the MPC's bid to extend the dam's license to 2015. In January 1989, FERC squashed that deal in its surprise decision to disallow an immediate extension of the dam license. The decision came as a result of the EPA's continued insistence that the pending FS, which could determine the dam's fate, took priority over the company's license. That would not be the end of the relicensing story, but never again would the company apply for a long-term license. Instead, "complete removal of the dam," the last possible remedial action (cleanup choice) buried in the EPA's "Initial Site Evaluation," would incrementally come to the fore, with the coalition leading the way.[29]

At the same time that the MPC lost its license bid because of concerns about the long-term future of toxins in the river, local residents began to attack the company's disposal of toxic sediments on land. As part of the Phase II permit agreement, the dam owners agreed to build an "upland disposal" site upriver from the dam and well out of the floodplain. All of the parties involved in the permit negotiations consented to that plan. But the five families that lived within a few thousand feet of the new disposal area weren't part of the planning party. In a letter to the MPC's vice president, those residents expressed their anger at being left out of the process, as well as about specific health hazards the site posed to their families. They complained of being "subjected to large volumes of dust for at least three weeks, twenty four hours a day." That dust, they worried, contained toxic metals from the reservoir. Even after the dust settled, the residents faced the economic loss of living next to a hazardous waste disposal site that was temporary but had no specified timetable for removal. Their list of financial losses included "property devaluation, loss of development potential and inability to obtain loans due to proximity of the dump." Their initial letter ended by lamenting that the disposal site infringed on the natural beauty that drew them to their current, and now sullied, homes.[30]

Two months later, when an EPA project manager responded to the homeowners' concerns, he failed to address the point that the Superfund process had left the most locally situated residents out of its decision making. Rather, he reaffirmed that the site was probably temporary, pending the final Milltown cleanup plan, which he incorrectly assumed "will be made within the next couple of years." Mostly he tried to assuage worry by summarizing the safety of the temporary site, constructed under EPA "guidance." Unsatisfied, the residents penned a more thorough letter, reiterating their initial concerns and then detailing specific problems with the disposal site that they observed during its construction and filling. These problems included possibly defective or compromised plastic lining in the disposal pit and sloppy handling of hazardous materials. The group closed this time with a battery of questions about responsibility for the site's construction and monitoring, as well as the demand that the EPA deem it temporary regardless of the long-term solution to the rest of the Milltown Reservoir. This second letter forced more than just a courteous response.[31]

Whereas local residents had famously taken two EPA agents hostage during the conflict over hazardous waste at Love Canal, resulting in the passage of the original Superfund legislation, it took only two letters from local homeowners to force a thorough EPA response at Milltown. Between the Love Canal story in the late 1970s and one of the largest Superfund projects in the country a decade later, the federal agency had learned to respond much more quickly to homeowners' questions, demands, and fears. In fact, as Milltown illustrates, Superfund made public participation a fundamental part of federal environmental cleanups. That transformation really happened only with the renewal of Superfund in 1986, not with the 1980 bill. Touting the new law's major changes in the *New York Times,* U.S. representative James Florio, who wrote the original Superfund law, put increased public participation at the top of that list. Whereas previously Superfund obligated the EPA to inform citizens of its decisions, the new law, in Florio's words, gave people "the right to be involved in the decisions." That is, Superfund provided the means to institutionalize the most grassroots environmentalism.[32]

With the backing of the Clark Fork Coalition, the state, the EPA, and eventually FERC, residents got the MPC to agree to build tree-lined earth berms to shield their homes from the disposal site. The dam owners also had to commit $50,000 to study the improvement of fish habitat in the river. This episode in the Milltown story illustrates how the focus of Superfund cleanup remained on mitigating immediate or temporary hazardous wastes in the course of finding permanent solutions.[33]

While the MPC came under fire for its temporary disposal site, a permanent fix for the Milltown site crept forward. By the summer of 1988, the reservoir's major responsible party, ARCO, had agreed to its PRP status in general and had begun participating in the process of completing a feasibility study. ARCO's consent was the ideal scenario intended by the recently revamped Superfund legislation. This "enforcement lead" option, in which the responsible party paid for work under EPA direction, saved time and money. ARCO also proved an ideal PRP because the company did not lack the latter. The company's first-quarter report for 1989 indicated a net income of $704 million, which was up more than 75 percent from the previous year. On top of those gains, ARCO had set aside $345 million, part of which was "for future environmental expenses." As one of America's top ten largest energy companies (mostly oil and gas), ARCO had the deep pockets that the Superfund law aimed to pick in order to pay for historic environmental degradation. That was especially important as Superfund neared its tenth birthday. Some of the strongest criticism about Superfund targeted the program's inability both to oversee timely cleanups and to make polluters pay. As an Associated Press story in 1989 pointed out, a recent Rand Institute study reported that of the 1,175 Superfund sites on the NPL, the EPA had declared only 34 fully cleaned in nearly a decade. Furthermore, corporate polluters had "paid less than a tenth of the cost" that Superfund had incurred during that time. With similar critiques of Superfund failures appearing regularly in the national press, laying responsibility on solvent and culpable companies like ARCO was important; making that responsibility stick was even more so.[34]

Perhaps ironically, ARCO's deep pockets had little to do with its holdings in the Clark Fork basin. When the company bought the Anaconda Company in 1977 it was part of a plan to diversify and grow. ARCO had recently expanded its oil drilling in Alaska and internationally and had increased its refining and marketing capacity in the United States. The company owned 21 percent of the Alaska Pipeline, which at a final price tag of $10 billion was the most expensive private construction project in U.S. history when it first transported oil from the continent's largest oil field in Prudhoe Bay, Alaska, in 1977. Executives hoped that the company could add copper mining to its portfolio with the acquisition of Butte mines and the smelter works in Anaconda, Montana. Under the leadership of CEO Robert O. Anderson, ARCO grew its oil business, but the Anaconda mining purchase proved a financial burden. By the time the EPA designated

Milltown one of its first Superfund sites in 1983, ARCO had abandoned all mining operations in Montana because of slumping copper prices and inexperience in the business. By 1986, new chief executive Lodwrick M. Cook began to pare the company's less profitable holdings. The strategy of concentrating on high-profit operations worked, with the major exception of ARCO's Montana properties. Superfund liability made it virtually impossible for the company to sell its shuttered mining and smelting operations, with the exception of some of Butte's open-pit mines. Montana entrepreneur Dennis Washington bought a few such pits for $18 million in 1985. He negotiated the purchase so that ARCO retained liability, and within a decade and a half he would turn nearly $1 billion in profit from the deal. Meanwhile, ARCO had to deal with the mess.[35]

During this same period, as the EPA encouraged state involvement, the agency decentralized its own authority. The 1986 SARA bill still contained clauses from the original Superfund law that vested authority in the U.S. president. In reality, after the passage of SARA, authority moved in the direction of the EPA regions. On most cleanup decisions, the president deferred to the EPA administrator, who in turn extended similar deference to the director of each EPA region. Similarly, more money went to regional offices following the passage of the new bill. Five years after Reagan signed SARA, the EPA was allocating 43 percent of its budget to regional offices, up from just 15 percent when the original Superfund law took effect. Through these delegations, Superfund oversight became more local. That was especially true for states like Montana that had a subregional headquarters. For Milltown that meant that the real seat of EPA power was the Region 8 office in Denver and the state office in Helena.[36]

Even as the EPA tried to shift the financial burden of a Superfund site to PRPs, the agency had to grapple with its older model of paying for actions out of the public coffers and collecting from corporations later. It wasn't until mid-1989 that the EPA took ARCO to court in an effort to recover some of the $6.3 million the agency had spent in its emergency responses at Milltown. This was different from suing the company. Rather, Superfund legislation allowed for "Judicial Consent Decrees" wherein a court supervised the reimbursement by PRPs to the EPA for work already done. It was a consensual contract. The company continued to agree to all of its responsibilities at Milltown.[37]

As the EPA shifted from its older pay-first-collect-later model toward its newer mandate to enforce work, the agency also began implementing ways to include more public participation, some of which originated with Milltown. During the summer of 1989, the EPA began holding meetings with a "Citizen's Advisory Group." The point of these meetings was to gather input from the public as well as to keep people informed of what was happening at Milltown. Superfund included a provision that allowed such groups to apply for grant money to help affected communities organize, distribute, and collect information. Through its meetings, the EPA helped the Milltown group apply for three-quarters of a million dollars for those purposes. What was unique and formative about the Milltown group was its request to observe "the upcoming negotiations with PRPs." Although the EPA denied local citizens that right, the agency did begin releasing summaries of all its dealings with PRPs to the public, through the advisory group. The decision to do so was, according to the EPA, "innovative" and marked the opening of a process that "used to be secretive." Similarly, the EPA's regional project manager endorsed additional innovative ways that his agency would include the public in the Superfund process. In a letter to the Clark Fork Coalition, Ken Wallace wrote, "EPA is committed to doing its endangerment assessment work in a public forum—something that EPA has not done before." Allowing the public to observe and comment on the agency's efforts to identify all the hazards posed by the Milltown Superfund site included designating a special Endangerment Assessment Committee. Wallace committed the EPA to funding the committee and "paid advertisements in the paper" in an effort to attract public input. His hope was that "full public involvement will become a routine part of the remainder of site work." This change marked one more way in which Milltown increased the importance and impact of public participation in national cleanups. Or, as Wallace noted, Milltown signaled "a shift in policy toward much greater public participation in Superfund."[38]

Once again, when it came to public participation, Milltown would reshape the way the process worked. And that reshaping would take far longer than the anticipated two years. In early 1990, the EPA wrote a press release encouraging the public to comment on ARCO's work plan for a feasibility study. The study was supposed to be a two-year endeavor, at the end of which the EPA would again solicit public comment on the cleanup options that were the centerpiece of any FS. The EPA considered the public's thoughts on the list of solutions the most important part of the process.[39]

Anticipating the two-year wait, the MPC filed a request with FERC for a six-year license extension. Forgoing a full relicensing, the company demanded no repayment by the EPA if the Superfund solution at Milltown Reservoir impacted the dam. In fact, in the permit request, the MPC recognized that the reservoir's designation had "elevated Milltown's status from an ordinary dam to an extraordinary dam." The company also used the opportunity to take a subtle jab at the federal law that kept thwarting its efforts, writing: "From amending its license, to reconstructing the dam, and finally, and perhaps most tellingly, to relicensing, MPC, state and federal agencies and the public have discovered the pervasive influence of CERCLA." Without acknowledging its own responsibility for the site or the FS, the MPC argued that even if ARCO and EPA finished the FS within two years (by 1992), the company's license would be on the verge of expiration and it would not have time to renew it. Thus, it was reasonable to give the company a short-term extension, which would run from 1993 to 1999, by which time the FS should have been completed and a solution at Milltown under way. Of those possible solutions, the MPC mentioned the complete removal of its dam. If that were to happen, it claimed, a short-term license made perfect sense.

No one at the time was advocating or seriously considering dam removal, yet the dam's owners were either prescient or including a logical, if unlikely, alternative, the opposite of which would be to do nothing with the dam. This new, short-term license and possible dam removal alternative met with more approval than almost anything the MPC had done since the issue of its dam repairs and licensing arose.[40]

~ ~ ~

The longer-than-expected timeline for finding a permanent fix to a major portion of the nation's largest Superfund complex had two significant benefits, both of which came into view during the MPC's early relicensing struggles. First, the methodical nature of coordinating multiple government agencies and provisions within Superfund helped provide the time and mechanisms for ever-greater public participation. As the EPA and ARCO deliberated over the FS and the dam owners crept through the permit process, local residents learned more and more about what was happening in and to their local watershed. Through press releases, newsletters, public meetings, and letter writing (most of which the EPA and Clark Fork Coalition promoted, if not funded), citizens digested the basics of the Superfund process, kept abreast of early cleanup work, and had (or created) opportunities to voice their opinions. In cases like the wetland, and then upland disposal sites that the MPC used, people's opinions altered what the company did with its toxic wastes.

In addition, the environment benefited from the ever-lengthening Superfund process. Public input also led to this second result. While Milltown inched toward a long-term cleanup plan, the Clark Fork experienced incremental improvements. As awareness of the fouled river grew, the city of Missoula banned harmful detergents and upgraded its sewage treatment plant. The MPC's license extensions increasingly included provisions for mitigating effects of the dam's operation on fish and wildlife. The company had to slow and more closely monitor its drawdowns. It funded fish passage studies and wetland development projects to make up for its disposal site gaffes. In these ways, Superfund's mandate to clean up places laid to waste by major industry bore proverbial fruit, such as a cleaner river, long before a list of possible remedies even emerged. That is, the process of being a Superfund site improved Milltown, even as the final product remained unclear.

What did become clear was that the Milltown Dam was going to be a major part of fixing the toxic reservoir, even if the dam's owners persisted in denying their liability. Indeed, the MPC may well have been the first to take seriously the idea of removing the dam, although it would eventually oppose that option just as it had objected to its financial responsibility during the early going. Almost as ironically, ARCO, the company that most readily accepted its liability at the outset and embraced the chance to study and write the initial feasibility study, would end up fighting the eventual solution therein. But before all that came to pass, floodwaters would once again alter the process at Milltown by threatening to destroy the dam, which raised public concern and participation. Figuring out a final solution at Milltown took another decade, another flood, and a fish, among other things. All of these, plus time, water, and wildlife made people look ever harder at the dam.

Nearly a decade after the passage of CERCLA, the new EPA administrator, William K. Reilly, admitted that "at the time of the program's initial enactment in 1980, virtually no one had any practical experience with the difficulty and duration associated with cleaning hazardous waste sites to health-based levels." SARA amendments had altered significant provisions of the law but had not increased that "practical experience." Thus, Reilly's first major project as administrator was to conduct a thorough examination of Superfund's first nine years and create "a realistic plan" for its future. Implementation of Superfund continued to emphasize "removal and remediation of immediate threats" to human and environmental health. Restoring environments was not part of the law's mandate or strategy for the future. At Milltown, it became so because of grassroots efforts that occurred outside of Superfund-mandated procedures.[41]

CHAPTER 4

Seeing the River As Resource

Suing for Damages

This is our one chance in history to bring this area back from a century of pollution.

—Montana Rep. Mike Kadas, D-Missoula, on House Bill 305 to
continue funding the state's lawsuit against ARCO (1995)

Shortly after the designation of Milltown as a Superfund site in 1983, George Ochenski went snorkeling down a stretch of the upper Clark Fork. Ochenski's outing was certainly an anomaly among popular, or even conceivable, uses of the river. It was less so in the context of the man's life. Ochenski had participated in the first ascent of the west face of Alaska's Mount Hayes; he had taught himself the art of cobbling mountaineering boots in an age of mass production; and he lived in a marginally restored cabin in a rather remote Montana ghost town. Yet Ochenski was no hermit. A freelance journalist and autodidactic environmental scientist, he also worked for the Montana Environmental Information Center during its campaign to help get Milltown on the EPA's National Priorities List in the early 1980s. Having published stories about the history of the industrial mining and smelting that had made a wreck of the Clark Fork, and warning people of the perils of ignoring that wreck, Ochenski knew what he was getting into when he donned mask, snorkel, fins, wetsuit, kneepads, and "a fanny pack holding three cans of beer."

In addition to knowing the river's history, Ochenski had river snorkeling experience. He had run a number of Rocky Mountain rivers, including a thirty-eight-mile pass down the Salmon River in Idaho and a whitewater-filled section of Montana's Madison River "famous for biting kayaks in half." Reflecting on the pummeling he took during that latter venture, Ochenski admitted he may have made "a miscalculation." Snorkeling the flat water of the upper Clark Fork below its Warm Springs tributary in Anaconda held a far different purpose for Ochenski. He hoped to see what lived in the river, rather than simply try to live through the experience himself.

After hearing about the Montana River Snorkelers Association and its membership of one, nature writer David Quammen decided to join Ochenski on one of his obscure outings. The article Quammen wrote for *Outside* magazine offered people a fish-eye view of western Montana's largest, and possibly deadest, river. Still, few people actually saw, as Quammen marveled, how "the rocks of the stream bed were largely cemented together with silt, leaving no habitat." Few looked in vain for a single fish or aquatic insect. Yet what the pair of snorkelers saw, or failed to see, would eventually help change the course of Superfund action along the Clark Fork.

While Ochenski's method of seeing the river was rare, many other people began to consider the overall health of the Clark Fork. Superfund required no one to stick his or her face directly into the river, but it did carry the requisite of research. Just as Ochenski turned his masked eyes directly on the ecological well-being, or poor being, of the river, others took on the task of looking at that same damaged ecology through other lenses.[1]

~ ~ ~

By the mid-1990s, a growing body of research explained that periodic floods had washed the toxic sediment from Butte mill tailings and Anaconda smelter works. Whereas the normal flow of the Clark Fork carried a trickle of the poisons, it was massive spring runoff that was responsible for piling heavy metal–laden silt behind the Milltown Dam and encrusting nearly 120 miles of floodplain. The flood of 1908 that nearly destroyed the dam likely carried a large load of the toxic sediment.

The same EPA-mandated study efforts also began to reveal a much more detailed picture of the damage more than a century of industrial mining had done to the Clark Fork and the Milltown Reservoir. Under the direction of the EPA and the watch of local river advocates like the Clark Fork Coalition, the Atlantic Richfield Company continued to pay for the bulk of this research. Corporate-funded, contract research proved how industrial pollution had harmed everything from waterborne bugs and earthworms to native fish and generations of people. And that damage was not an inert, historic fact. River users continued to lament the degraded condition of the watershed. Following intense thunderstorms, schools of belly-up fish floated down the Clark Fork and into the public eye and ire.

The attention and information on how toxic waste had imperiled aquatic resources, and continued to do so, amplified the first, largest, and most contested lawsuit of the Superfund process at Milltown. The same year Ochenski snorkeled the nearly barren river, the state government implemented a Superfund provision to sue ARCO for damages that western Montana's legacy of industrialization had inflicted on natural resources. Trying to recoup the loss of natural resources plunged state agencies and the major responsible party into a long and contentious legal battle, the settlement of which would set the parameters and provide the funding for the transition from containing toxic waste to restoring the environment. But it was the attention on overall Clark Fork ecology, which spiked following the death of a few thousand fish in 1989, that transformed the state's suit from a rather small endeavor to a major restorative component of the river's Superfund history. Actual restoration work was a long way off, but its origins lay in the ability of the state to capitalize on natural resource damage provisions within Superfund.

Detailed study of river health on the Clark Fork came out of a larger transition within the EPA toward streamlining the federal government's efforts to address perennially strong public concern for the environment. As criticism mounted about the wastefulness of the nation's signature toxic waste cleanup program, the EPA tried to rationalize Superfund law, emphasizing better scientific research, clearer cleanup standards, and a keener eye for cost-benefit analysis. The case of Milltown illustrated the implementation of all three of these new national priorities for Superfund.

~ ~ ~

ARCO continued to craft the Superfund-mandated feasibility study, which would ultimately spell out permanent solution options at Milltown while simultaneously combating the state's efforts to recover damages for the historic destruction of its natural resources. On the one hand, ARCO was fulfilling the role of cooperative corporate partner that was the ideal of Superfund. On the other hand, the company embroiled the state of Montana in the kind of litigation that Superfund critics claimed was the legislation's weak point.[2]

In the public eye, the company's villainous side emerged in the summer of 1989 with the headline "Government Sues ARCO." The *Missoulian's* report actually dealt with the EPA's efforts to get the company to reimburse Superfund for some of the emergency work already accomplished along the Clark Fork, especially the new Milltown water supply, completed in 1985. The federal government's $5.7 million claim was for remediation, which Superfund defined as a host of possible actions "to prevent or minimize the release of hazardous substances so that they do not migrate to cause substantial danger to present or future public health or welfare or the environment." Essentially, this meant containing wastes, or curing the symptoms therefrom. That was the essence of Superfund as it was first written. Quick emergency actions, such as replacing the drinking water supply at Milltown, were also "the unsung success story of the Superfund program," according to a thorough analysis of the program by the Brookings Institution. Dwarfing the initial federal claim, what the state asked for pointed toward a new and different goal of Superfund.[3]

The federal suit pressed ARCO for money it had already spent; now the state wanted money for future expenditures. Specifically, the state Department of Health and Environmental Sciences (DHES) had filed a claim against ARCO back in 1983, "seeking $50 million in compensation for damage to natural resources." That was the maximum amount the original Superfund law allowed states to claim against a PRP as compensation for damaged environments. The new Superfund law passed in 1986 removed that cap and liberalized the ability of federal, state, and local governments or individuals to pursue such claims; it also emphasized that the money was for restoration. The DHES's remunerative demands would soon increase, ARCO would fight the case for years, and the issue would come to encompass far more than the dead fish that touched it off.[4]

Industrial runoff had killed fish in the Clark Fork for a century by the time the state sued ARCO. But it was only after designation of the river as a Superfund site that the exact causes and blame for the deaths became well known. In its suit the state claimed that since at least the 1860s, ARCO's "predecessors-in-interest" had injured natural resources, including "fish, wildlife, surface water, groundwater, soil, and vegetation."[5]

Fish-killing events amplified general concern about the loss of the state's natural resources. "It was Armageddon revisited," dramatized the *Missoulian* in reference to a thunderstorm that washed enough old toxins off the riverbanks and into the Clark Fork to kill "thousands of fish" along at least eighteen miles of the waterway in July 1989. As state FWP biologists tried to assess the exact damages, the CFC's Peter Nielsen recalled that at least seven documented disasters of identical nature had occurred since 1981, when Milltown had first come to the EPA's attention.

Ire turned on ARCO. As the local newspaper reported, "area sportsmen said they were angry that all the federal and state studies have yet to result in any action to protect the fishery." The truth was that ARCO had been working to build a "temporary diversion dam" meant to channel rainwater rushing over especially contaminated sections of the upper river into settling ponds. The dam and the ponds were meant as a "Band-Aid" to prevent or lessen acute poisoning of the fishery until more thorough and permanent remediation work commenced, but spring high water had prevented progress on their building.[6]

Following the fish kill of 1989, letters from University of Montana scientists, the CFC, and riverside residents addressed to the EPA and the DHES lamented the recurring problem and pressed the agencies for a more permanent solution. These letters also emphasized how major fish kills were poignant symptoms of a much bigger problem: a lethally degraded watershed. As Nielsen put it, aside from periodic acute killings, the river suffered from "this low-level leaching," which left the fishery with a fraction of the fish it could support. None of this was a revelation but it lent support to the state's actions against ARCO.[7]

After the 1989 fish kill, ARCO offered to build more berms around toxin-laden stretches of the upper river and said it would contribute $1 million to state coffers "for use in mitigating environmental problems affecting the Clark Fork River." Attached to the funds was a request that the money be in lieu of further state action on the matter of the recent fish kill, such as the state's $50 million suit. In contrast, the state regarded the fish kill incidents as evidence of a much bigger problem, which deserved much greater funding.

Under the Superfund provision that empowered governments to sue PRPs for losses of natural resources, Montana hoped to restore the upper Clark Fork to its "pre-European condition." The law's section on "natural resources liability" specifically allowed claimants to pursue money "to restore, replace, or acquire the equivalent of such natural resources" as had been damaged or destroyed. The law offered no definition of what restoration meant. Aware of the magnitude of taking on ARCO, by the fall of 1990 the governor's office had hired a natural resources damage coordinator and appointed a special task force, the Natural Resource Damage Litigation Program (NRDLP), to oversee the suit. As for how the state might define restoration, CFC scientist Phil Tourangeau speculated that "in the best of all worlds the Clark Fork would return to its pre-mining state. Realistically, I don't know if that will ever happen." That work certainly wasn't going to happen with $50 million. When the state's NRDLP suit made it into a courtroom in 1997, its price tag had increased exponentially.[8]

~ ~ ~

As the state slowly moved forward with its suit, the federal government proceeded at a similar pace with its oversight of remediating the Clark Fork. Specifically, the EPA and ARCO continued to work toward producing a feasibility study, on which the fate of the Milltown Reservoir hinged. One of the main reasons it took so long to produce this important document that listed possible solutions for a Superfund site was the ongoing concern about public participation.

More than any other group or individual, the Clark Fork Coalition pushed the issue of how, when, and to what degree the public participated in the process. In addition to keeping an eye on the recent fish kill and counseling the state on its lawsuit, CFC director Peter Nielsen repeatedly requested that the EPA include more local input, or at least define the parameters of that input. Shortly after summer solstice 1989, Region 8 EPA director John Wardell wrote to Nielsen assuring him that the coalition's and other local views of the FS work plan were "being factored in to the final draft." Wardell also offered that public input would precede "negotiations with PRPs." In other words, the EPA planned to consider people's thoughts before those of ARCO. Even though federal and court regulations disallowed the public or the CFC from being part of negotiations with PRPs, Wardell wrote that "CERCLA regulations and common sense demand that we do more, by providing information to the public, and by providing for meaningful public input throughout the process." In more congenial language, the regional director also assured Nielsen that "if you have specific suggestions on how to improve this situation, I would like to discuss them with you." Recognizing the importance of the coalition as a window into public sentiment and a medium of good public relations,

Wardell wrote that "any help in this effort by your organization would be appreciated." The CFC's help, along with public input, left its imprint on both the process and the final product.[9]

Superfund mandated many of the ways in which the EPA engaged the public, but the reality of public input was not wholly governed by codification. Describing how written law affected reality, one Region 8 administrator commented that "here in the Regions is where the rubber hits the road." She went on to explain that it was in local and regional EPA offices, at public meetings, on job sites, and in local libraries or other public venues that citizens approached federal officials wanting to know whether or not a site was going to give everyone cancer or produce babies full of birth defects. In other words, most people understood Superfund through the lens of very personal concerns. Superfund provided a venue in which people could express their concerns and personnel at whom to direct them.

The EPA realized from early on in Superfund implementation that harnessing, even institutionalizing, public input was essential. In addition to publishing frequent articles about the topic, the *EPA Journal* devoted an entire issue in 1985 to "EPA and the Community." In his introduction, EPA director Lee M. Thomas claimed, "It is possible that we attract more intense interest—in the form of letters, phone calls, and attendance at local hearings—than any other agency of the federal government." Hence, the agency needed to manage that interest. While acknowledging that most interest in Superfund was very local and "grassroots"—not entwined with national environmental groups, Superfund community relations experts wrote that "a high level of community involvement in a hazardous waste problem and the development of opposition to government plans are usually linked to the way citizens believe they have been treated by the government." So, part of the EPA's goal was public relations, providing clear, accurate, and timely information to citizens. But "the program aims at two-way communication," according to the experts. In fact, the EPA ideal for Superfund was a conflict-free process in which "highly noticeable modifications are unlikely because planning can be continuously responsive to community needs." Citing numerous examples, EPA experts emphasized the point that citizen input could change Superfund actions, especially during the writing and debate over the feasibility study.[10]

People around Milltown directly benefited from the EPA effort to increase public involvement. In fact, Superfund encouraged new forms of environmental activism. The Superfund legislation passed in 1986 funded communities to form technical advisory committees (TACs). These TACs could receive up to $50,000 per year. As an EPA news release about the formation and funding of the Milltown TAC explained, the money allowed citizen representatives to "hire an independent technical advisor to help them understand and comment on technical factors in cleanup decisions affecting the Milltown and Missoula communities." These grants were another way that Superfund institutionalized grassroots environmentalism. The Milltown grant disallowed recipients from using its allotted funds for organizing citizens. Rather, it was explicitly for informing the committee and encouraging its participation in EPA-designated forums. The groups would join a host of other "civic and conservation groups" under the umbrella of the Milltown EPA Superfund Site committee (MESS). MESS, started as a citizens' committee by the CFC in 1986, illustrated the diversity present even in local, grassroots organizations. It also provided an example of how local environmentalism was not a reaction to or against national environmental politics in the 1980s, but at times a codified and supported part of them. Superfund encouraged new environmentalists.

MESS chairperson Tina Reinicke-Schmaus knew nothing about the EPA, Superfund, or any of the agencies involved at Milltown until she began wondering about the backhoes and trucks rumbling through her neighborhood on the southeast side of the reservoir. Upon learning that the machines were creating an "upland disposal site," she got involved. Writing in the group's newsletter, she attested that "members . . . came from diverse backgrounds: housewives, homeowners, the state fish and wildlife department, Missoula City/County Health Department . . . Trout Unlimited, MontPIRG, the Clark Fork Coalition and the Milltown Water Users." In addition to being socially diverse, the group dealt with many issues stemming from Milltown, rather than maintaining the single-issue focus attributed to most grassroots efforts. Early on, MESS was charged with working alongside the EPA, the state, and PRPs on various environmental risk assessments of the watershed.[11]

Risk assessments were a step toward a completed feasibility study. Meetings and comment periods ran through most of 1991 in preparation for an FS due in 1992–1993. Many of the comments addressed the various risk assessment reports that the FS required. During a spring 1991 comment period, the Missoula Water Quality Advisory Group, which consisted of "hydrologists, soil scientists, water chemists, engineers and others involved in water quality related professions," addressed each of the initial eleven recommended remedies for Milltown. In general the groups thought that "clean-up at Milltown without elimination of the upstream source would be foolish." These professionals recognized the interconnectedness of the river's ecology, and hence the need to let the flow of the river (and the toxins it transported) dictate cleanup order. The Missoula City-County Health Department offered a similar critique. The department took the long view of ecology, criticizing cleanup options that were temporary or that would allow similar pollution problems in the future, giving some of the first serious consideration to dam removal, and ending by declaring, "Resource protection and the logistics of a perpetual clean up plan should dictate the solution. Not decades, but centuries of habitation are envisioned for the Clark Fork River Basin. Ecological destruction in the last century is a click on the clock compared to future use of the valley." The Montana chapter of the American Fisheries Society touted its membership of "over 100 professional biologists" before commenting that, among other things, the initial reports about remedial alternatives at Milltown ignored "impacts on the aquatic ecosystem."[12]

These concerns about overall environmental health wove together Superfund's mandate to remove toxins and the emerging emphasis on fixing a broken ecosystem. It also fit into what a *New York Times* article, following the passage of SARA in 1986, deemed a movement toward thinking about the quality of Superfund cleanups. The revamped Superfund law included a provision that required the EPA to meet the most stringent environmental quality standards for each project, whether federal, state, or local. As a 1990 retrospective published in the *EPA Journal* summarized the law, "Congress directed EPA to focus on permanent remedies for Superfund sites rather than simply containing untreated wastes on site." Hence, cost-effectiveness and protecting public health no longer drove cleanup standards; environmental quality supposedly did.[13]

By the mid-1990s Congress was again tinkering with Superfund. In particular, Carol Browner, the EPA director under President Bill Clinton, favored a move toward what she called "different levels of clean." Superfund waste cleanup standards should match the projected future use of a site, according to Browner. Current or proposed residential areas would necessitate much cleaner solutions than prospective industrial sites, for example. At Milltown and along the whole of the Clark Fork the emphasis remained on turning the destructive legacy of industry into a thriving watershed measured by baseline environmental health.[14]

Having seen the damaged river up close through his plastic mask, George Ochenski summed up the emphasis on natural resource problems best. His "two cents worth" noted how providing new drinking water in Milltown "was merely addressing the *symptoms* rather than the *cause* of the metals pollution." Ochenski deemed the proposed remedies for Milltown to be symptom relief rather than causal cure. In contrast, he proposed "actions necessary to return this river to a healthy and productive waterway." In fact, Ochenski's list of incremental steps for fixing the Milltown problem was eerily prescient. His first step was attending to upriver cleanup and then to a slow lowering of the reservoir. Then came taking care of sediment removal, followed by dam removal, and finally, restoring the area to a "river-side 'confluence' park" run by the state. With this progression Ochenski laid out the basics of Milltown's future. He even made one of the key arguments that later came to be an incontestable part of the plan: "the power generating capacity of the Milltown Dam is strictly ancillary to the main issue of a permanent clean-up." Ochenski saw that arguments for dam removal and ecological restoration had to hinge on more than just a healthy environment. In conclusion he wrote, "These solutions provide *jobs* for working Montanans *restoring* their environment. . . . Many bright, young Montanans will dedicate their careers to restoring Montana's largest volume river to a clean and healthy state." Ochenski had company in thinking about the economics of Superfund.[15]

~ ~ ~

At the national level, criticism of Superfund mounted as the legislation faced another round of renewal. In his first State of the Union address, Clinton echoed calls for the EPA to rationalize its programs, Superfund specifically. By this the president and an increasing number of experts and politicians meant using scientific risk assessments to determine when and where to spend money on the environment, to create clearer national standards for cleanup work, and to heed cost-benefit analyses more closely. These goals and Clinton's appointment of Carol M. Browner as the new EPA administrator signaled a turning point.[16]

In 1990, during the George H. W. Bush administration, Congress had renewed Superfund with virtually no changes. That did not mean that legislators, or the public, saw the law as flawless. In a 1991 letter to the *New York Times*, the EPA's top administrator of emergency and remedial responses, Henry Longest, dispelled common criticism about his agency's most well-known program by rebutting claims that Superfund was failing to attend to the nation's toxic wastes. Though critics routinely pointed to the slow increase in NPL sites, Longest pointed out that while the EPA had only 1,200 designated sites, the agency had screened 33,000 more, of which "21,000 have been determined *not* to require long-term remediation." That is, the EPA was aggressively assessing the country's pollution problem. To the charges that Superfund was financially bloated and its implementation misspent government dollars, he reminded readers that Superfund survived on taxes collected from industrial corporations, not the public dime.[17]

Taking the reins of the EPA, Carol Browner was well aware of the calls for change that the agency faced. She knew that reforming Superfund had to be part of that change, since the program accounted for one-fourth of the agency's yearly $7 billion budget. To the task of reform she brought a new type of leadership that featured a personal love of the environment.

Browner grew up exploring the Florida Everglades. After earning English and law degrees, she helped engineer the nation's largest ecological restoration project in the stomping ground of her youth. Upon accepting the EPA's top position, she described her motivations as wanting her "son to be able to grow up and enjoy the natural wonders of the United States in the same way that I have." Browner combined her passion for the environment with her proven business savvy. As a *New York Times* story announcing her appointment put it, "Ms. Browner is prominent among a new type of environmentalist who views economic development and environmental protection as compatible goals." In the same article, Browner addressed her priorities of upholding the public desire to improve the environment and get business on board, saying, "I've found business leaders don't oppose strong environmental programs. What drives them crazy is a lack of certainty. We can change that."[18]

The changes in store for the EPA and Superfund may have come out of personal environmental passions such as Browner's, or the kind the public expressed in response to national catastrophes, but those changes meant to favor science over fervor. Discussing the EPA's history of addressing environmental issues, a 1993 *Times* article reported that "experts say that in the last 15 years environmental policy has too often evolved largely in reaction to popular panics, not in response to sound scientific analyses of which environmental hazards present the greatest risks." William K. Reilly, one of Browner's predecessors at the EPA, called the problem "agenda-setting by episodic panic." Reilly pointed to how the creation of and changes to national environmental policy had hinged on public outcry over "Love Canal, Valley of the Drums, [and] the Exxon Valdez." Browner agreed with Reilly's (and most experts') notion that environmental policy ought to come from persistent problems and sound science. The *Times* dubbed the use of risk assessment and prioritizing through rational, science-based decisions and cost-benefit analysis "the vanguard of a new, third wave of environmentalism that is sweeping across America."[19]

It may have been an exaggeration to say that a new environmentalism was "sweeping" the country, but changes were in the making. Between 1993 and 1995, four of the nation's most important federal environmental policies were up for renewal, including Superfund. The focus of the new Democratic policy makers was on making those laws more methodical and less reactionary. As the 1995 Superfund rewrite approached, those arguments peaked. Browner led the EPA in recommending that Congress amend Superfund to fit with the agency's new approach of favoring risk assessments, that is, letting scientific studies of environmental dangers dictate the range of possible solutions. The EPA was also pressing for clearer national standards for environmental cleanup based on a site's potential future use. Conversely, Congress tried to relax regulations on polluters and trim the EPA budget. Under White House pressure, congressional Republicans dropped their efforts to remake Superfund into a basic waste containment program, or what one D.C. environmental leader called "Superfence." A commentary in the *EPA Journal* explained the Clinton administration's clear victory over a heavily Republican Congress elected in 1994 as a result of the environment continuing to garner overwhelming public support. Exit polls showed that the same people who ousted Democrats from national and state legislatures in 1994 still supported tougher environmental laws, not weaker ones. And they, like Browner, thought that the environment and the economy were not a zero-sum game.[20]

In spite of all the criticism, studies showed that Superfund was leading to swifter, more cost-efficient cleanups. In 1987, after the passage of SARA, responsible parties led only 39 percent of remediation projects. By 1993 that had risen to 79 percent because of the EPA's implementation of SARA. ARCO's willingness to accept the responsibility of researching and writing the Milltown FS was a case in point. The same studies that tracked the rise in the percentage of companies leading Superfund projects also indicated that that strategy made remediation cheaper and more rational. Corporate cleanup with federal oversight seemed to be working. In the fall of 1995, moderate Republicans broke with their more conservative party members and joined Democrats in pushing a Superfund renewal bill that left corporate liability intact and strengthened the law's emphasis on corporate-led, scientifically driven waste cleanups.[21]

Yet by early 1996, Superfund was on the verge of a temporary shutdown. In late January 1996 Carol Browner testified in Congress about how the budget impasse was damaging her agency's most vital programs, including Superfund. She admonished the legislative body over deeper proposed budget cuts that would redouble that damage. While not claiming a premature victory, Browner and D.C. environmentalists sensed that the GOP was backing away from its fiscal hacking of federal environmental programs. Some even saw those programs as approaching the status of issues such as Social Security, which were practically immune to congressional cutting because "they carry such voltage with voters," according to the *Times*. By early spring, Clinton even toured a few Superfund sites that Congress's budget struggle had halted. Vowing to veto any legislation that cut Superfund, Clinton blasted conservative Republicans for trying to "let the polluters off the hook and make the taxpayers pay."[22]

In April, Congress restored the EPA's coffers to nearly their previous level of just under $7 billion. Superfund remained fiscally intact, but Congress still had to pass another five-year extension of the bill. Even House Speaker Gingrich's vision of a "new environmentalism" sounded very similar to the changes the EPA was already pushing. His notion of a renewed Superfund included "valid science, incentives for compliance rather than punishment for noncompliance, rapid adoption of new technologies, and a search for innovative solutions from communities." With the exception of replacing the compliance stick with a carrot, he sounded much like Browner, and a lot like what was already happening on the ground at Superfund sites.[23]

Just as the spring of 1996 appeared to be another watershed moment for Superfund, it was perhaps the most important turning point in the law's implementation at the Milltown Dam. By the time Congress restored the EPA and Superfund budgets, the Clark Fork had once again reshaped the view of Milltown. Before that reshaping happened people were already questioning the path that the EPA and ARCO were setting for the site.

~ ~ ~

In the early 1990s, river advocacy groups such as the Clark Fork Coalition and the locally organized Milltown EPA Superfund Site committee

(MESS) both hinted that the dam did not belong in a long-term solution. Specifically, the groups challenged the idea of containment—keeping sediments in place—because it hindered the long-term prospect for restoring a healthy ecosystem. As MESS scientist Phil Tourganeau put it, the emphasis "should be the permanent (in geologic time) cleanup of contaminated sediment and groundwater to the degree that public and environmental health is unequivocally protected." Throughout these comments groups aligned themselves with local citizens. As one individual wrote, "Keep in mind that the public has the greatest stake in this cleanup. We live here!" The dam owners also tried to align themselves with locals. The MPC supported dam engineering remedies and institutional controls for the Milltown problem, claiming that the dam was "operated and maintained for the customer's benefit." The company went on to identify its ownership of the dam with ecological resources such as "waterfowl and deer hunting," for which the MPC had historically opened its private land. Furthermore, the company promoted an extension of the trail network on its land around the reservoir as its concern for encouraging a healthy environment. Its hopes for keeping the dam intact and improving it hinged on the language of an improved environment, as was true for proponents of more aggressive cleanup measures.[24]

Not all the attention to the current and future health of the watershed came via public comment or its opposition. Drafting a feasibility study meant researching the river system. In an *EPA Journal* article about the agency's regulatory capacity, the authors asserted, "All EPA's decisions must be based on sound science."[25] Private contractors and state environmental agencies began to produce studies on everything from aquatic insects and earthworms to waterfowl and fish. These studies, paid for by the EPA and ARCO, factored into the FS, as well as the state's suit. Study after study completed during the early 1990s determined that heavy metals from the basin's mining history had damaged the river's natural resources. Soil-eating worms accumulated high levels of heavy metals, which they potentially passed up the food chain when preyed upon by birds or fish. Macroinvertebrates, called "benthos," proved full of arsenic and other heavy metals as well. These spineless aquatic creatures, including crayfish, snails, and more worms, as well as a host of insects that spend most of their life cycle clinging to submerged river rocks before they hatch into their winged stage (giving them names such as stonefly and mayfly), are the staple of fish diets. People who fly-fish cast imitations of many benthos to entice hungry trout. So native fish also tested exceedingly high for arsenic, copper, and zinc—at least those that researchers could find. Studies showed that similar Rocky Mountain rivers supported thousands of trout per mile, while populations in the Clark Fork from its headwaters to Milltown languished at about a quarter of the expected numbers. Many of the surviving fish showed inhibited growth that correlated with a toxic diet, while others had abnormal scale loss or irritated gills clogged with mucus. ARCO challenged some of the growing body of research (that the company itself funded) on the river ecosystem, in one instance contesting that toxic metals were the most likely cause of the degraded fishery.[26]

Alternatively, the Montana Power Company supported the scholarship pointing to a river degraded by historic contamination. As part of its license extension, the company had to address the effects the dam had on the river's fishery and propose a "mitigation and enhancement plan." Its assessment emphasized the damage resulting from toxic sediment and downplayed the role of the dam in blocking fish migration and spawning. The company proposed slower drawdowns of the reservoir to decrease fish-harming sediment flushes, as well as $65,000 per year to help restore tributary streams important for fish spawning. It would also breed and release hatchery fish and monitor the effects.[27]

Whereas most of the research on a collection of organisms that defined the river ecosystem focused on the historic and contemporary effects of contamination, most of the concern for human health looked forward. Reports by a Public Health Working Group working at Milltown evaluated "sources of exposures to humans visiting the reservoir." The primary concern seemed to be "projected recreational use of the reservoir." Identifying these pathways of human exposure, such as where families got their wild game, whether or not they ate vegetables from a home garden, or how often they swam in the reservoir, would help direct future cleanup that in turn would reduce future exposure. When it came out in July 1993, the EPA-mandated "Baseline Human Health Risk Assessment" for Milltown reported results from surveys of Milltown residents meant to determine the pathways by which residents might have, did, or could expose themselves to toxic sediment and water. The Montana Department of Health and Environmental Sciences praised the report for its thoroughness and the "unusually open forum" in which the EPA produced it, including participation by the locally organized MESS.[28]

It wasn't until 1993 that an initiative to look at historic health effects in or near Milltown arose. And it arose through local public input. In a letter written on Halloween 1993, Barbara Bennetts wrote to a member of MTAC asking about what the group knew or was doing about reports of "high incidents of cancer among families living along the river." Bennetts included in her concern the spooky anecdote that "there is one cluster of homes . . . six or seven in all . . . where one hundred percent of the households have a cancer victim." Before the end of the year the EPA recruited staff from the Centers for Disease Control's Agency for Toxic Substances and Disease Registry (ATSDR) and the state DHES, "requesting an investigation into the question of potentially high rates of cancer around the Milltown Reservoir." The EPA manager involved, Julie DalSoglio, also accepted Barbara Bennetts's offer to act as liaison between agency people and community members. But as DalSoglio explained, the impending study by the DHES and ATSDR would be of the state's tumor registry, which included hospital-treated cases of cancer. The registry lacked information on cases treated outside a hospital or outside the state, as well as on some specific types of cancer, like that of the skin, which is the primary risk associated with ingesting toxic quantities of waterborne arsenic. The Missoula City-County Health Department asked that the DHES keep it informed of any cancer cluster studies to come. Meanwhile, one of the most important studies of the site that assessed human health risks dismissed any significant worry about cancer, except in the case of people drinking groundwater. And since no one was doing so anymore, the report deemed that risk low. Surprisingly, the report said nothing about the historic effects on the people who had consumed groundwater on a daily basis for years, if not lifetimes.[29]

The EPA organized a review of cancer in and around Milltown. At a May meeting in 1994, the agency got representatives from both the ATSDR and Montana's DHES to agree to initiate a cancer cluster review for the area. They hoped to supplement the state tumor registry with information collected from residents' medical records, if they could get voluntary participation in the study. Over a year later, in July 1995, the ATSDR epidemiologist on the case reported that there was "no data to show cancers in the area . . . outside normal, expected values." In a letter to Barbara Bennetts, who had first broached the subject with the EPA, Julie DalSoglio stated that in the end the study included only "a review of death certificates." The letter admitted that the results failed to fully address Bennetts's concerns and it put the onus back on her to figure out additional funding for a more thorough study by the ATSDR, funding that the EPA could not provide.[30]

~ ~ ~

Despite the completion of even more studies, the EPA considered deferring the feasibility study of Milltown. In a July 1992 newsletter coproduced by the EPA and the DHES, the agencies publicly announced the possibility of delaying the FS at Milltown "largely because the sources of contamination upstream have not been addressed." Like so much in the Superfund process, the decision to delay producing this vital document by carving the reservoir and dam out of the rest of the Clark Fork plans included consultation with local advocacy groups. Even before the public announcement, the new director of the CFC and Peter Nielsen, who had just taken a job with the Missoula City-County Health Department, questioned the EPA about the possibility. Everyone, including ARCO, seemed to agree about the wisdom of deferring Milltown to allow more upstream remediation to progress. But that wasn't the only justification.[31]

By the 1990s there was an emphasis within Superfund on streamlining cleanup efforts as a way of making them more fiscally efficient. A 1990 edition of the *EPA Journal*, a report entitled "Progress and Challenges: Looking at EPA Today," recognized that "the number of abandoned hazardous waste sites has turned out to be much larger than was predicted when Superfund was created. Furthermore, cleanup has turned out to be complex, taking more time and resources than expected to complete the job." In 1982 the EPA listed 418 Superfund sites, including Milltown. By 1990 the agency had more than 1,200 active sites, compared to 52 completed cleanups. Milltown was representative of what was happening all over the country. The vast majority of Superfund sites took decades to remedy. Most of the "time and resources" were chargeable to the PRP, but the EPA still had to conduct each Superfund project, and it added more every year on a limited staff. At a January 1993 public forum held in Missoula, EPA project manager Julie DalSoglio recognized that limited staffing was a partial cause of the delay in the Milltown process. The agency considered delaying work at Milltown to be a way to shift limited resources to upriver cleanup. At the same time, the final designation of Missoula's aquifer as a "sole-source" water supply lent weight to the wisdom of fixing upstream problems prior to assessing the future of the reservoir and dam, which kept the bulk of the watershed's contamination out of Missoula's water. A month later the MTAC newsletter confirmed the likelihood of the deferral, adding that the decision could delay work at Milltown for "3–5 years, with actual cleanup more years in the future." Unbeknown to anyone, the longer timeframe for Milltown would allow events to unfold that drastically changed the final outcome there. Like so much of the Superfund process at Milltown, end results hinged on unintended consequences.[32]

Many places across the West were now beginning to call attention to the unintended yet harmful consequences of the region's mining history. Milltown was affected by mining far from the tunnels and pits that Butte mining companies dug in search of copper. Plenty of other western towns were beginning to deal with the effects of mining much closer to their sources. A 1993 *New York Times* article estimated that water running over old mine tailings had polluted approximately "10,000 miles of streams in the West." All those stretches of degraded waterways added to the "staggering environmental cost from industrial development in the United States." Just as Milltown had harbored the poisons from mining the copper that electrified the country, countless other western environments had borne the brunt of material and technological advancement in America. According to the *Times*, the West's tolerance of that legacy was waning and communities were beginning to demand studies and remedies for their local environments.[33]

With a ballooning number of possible toxic waste sites to deal with, the EPA had to prioritize. Part of prioritizing entailed making risk assessments. In 1990 the EPA's Science Advisory Board endorsed a report called "Reducing Risk," which argued that the agency and Superfund, in particular, ought to focus on identifying the worst of the worst environmental sites in the country and then commit the most resources and immediate effort to remediating those places. As William K. Reilly, who served as EPA administrator for George H. W. Bush (and in 2010 became Barack Obama's point man for investigating the BP Deepwater Horizon oil spill in the Gulf of Mexico), summarized it for the *EPA Journal* shortly after the report's release, "Setting environmental priorities for the whole nation and bringing our Agency's resources into alignment with those priorities are supremely daunting tasks." The way to accomplish those tasks was to prioritize via risk assessment, rather than being "swayed by the passions of the moment," Reilly surmised. And risk assessment meant using "sound science." In other words, the EPA's top administrator was saying that no longer would the squeakiest wheel get the grease. The keystones of defining risk remained, as Superfund law prescribed, human exposure pathways and toxicity of pollutants. In 1993, by the time the deferral of Milltown's FS became an issue, risk assessment was a well-established practice in the EPA's implementation of CERCLA.[34]

At Milltown the flow of the river added to the discussion of risk assessment. With arsenic in groundwater at issue, the Milltown portion of the Clark Fork Superfund complex scored high for risk, but cleaning from upstream downward made a great deal of sense, especially since the residents were no longer drinking from the contaminated wells in Milltown. Yet public opinion trumped these other considerations. In autumn 1993, John Wardell, the EPA's state manager, announced that Milltown would remain a priority within the river-wide Superfund complex and that work toward completing an FS would continue. The CFC praised the decision and the agency's "willingness to listen to the concerns of the Missoula community. We believe that your decision is a solid one and one that is supported wholeheartedly by Missoula."

Others agreed. The Milltown Technical Advisory Committee began its first newsletter of 1994 commending the decision. The county health department praised the EPA and emphasized recent studies that indicated that arsenic-tainted water could move from the Milltown aquifer into the Missoula aquifer. With "as much as 35% of the recharge to the Missoula Valley Sole Source Aquifer com[ing] from Milltown," the health department claimed, "public interest in the remediation of the Milltown groundwater site is very high in this community."[35]

ARCO, on the other hand, was quite a bit less happy. The PRP complained that it had spent considerable money and time proceeding with the upriver FS under the impression that the EPA planned to defer the Milltown FS. In addition to being surprised and disgruntled with the agency's choice not to defer on Milltown, a company spokesperson wrote that the "decision was communicated to special interest groups either concurrently or before [the] call to me . . . which has done little to engender a cooperative relationship between our organizations." Even though ARCO opposed the decision and the EPA's apparent favoring of local advocacy groups, the studies and writing for the FS crept forward, as did the state's lawsuit for historic damage to its natural resources.[36]

~ ~ ~

The state of Montana intended to use many of the same studies being conducted for the EPA's FS to bolster its natural resource damage claim against ARCO. The attention to and evidence about how contamination had degraded the Clark Fork watershed emboldened the state. By 1993 the state had contracted Hagler Bailly, an international consulting firm that, among other things, specialized in natural resource damage claims. In a 1993 report, the consultant's scientists reported that the "Clark Fork River Basin contain[ed] some 9,900 tons of arsenic . . . 99,000 tons of copper, 22,000 tons of lead, and 55,000 tons of zinc." Those contaminants had led to a nearly lifeless river at times during the last hundred years, had repeatedly caused catastrophic fish kills in the last decade, and now persisted in degrading nearly all forms of aquatic life in the watershed, as well as posing an ongoing threat to human health. Hagler Bailly promised "to assess the monetary value of the injuries, and to establish a framework for the restoration of [the] resource" within the coming months. But the state's money and plans for restoration hinged on what remediation looked like. Finally, remediation, or the EPA decision as to how to deal with the contamination at Milltown, seemed to come into focus near the end of 1994, at least temporarily.[37]

In the fall of 1994, the EPA released a remedial investigation (RI), which was the culmination of studies done at Milltown. It soon followed with the first FS. Both were drafts, but they gave the public its first official glimpse of what might happen at the site. Titan Environmental Corporation, which wrote the drafts; ARCO, which paid for them; and EPA, which approved them, had narrowed twenty-three possible cleanup solutions to six. The winnowing process had involved "evaluating short- and long-term aspects of effectiveness, implementability, and cost," according to the drafts. The extremes of the original, longer list included everything from "no further action" to "Dam Engineering (Dam Removal)," which scored minor points for long-term effectiveness.[38]

The first of what would be multiple "final" reports came out in early 1995. While it acknowledged that the river was flushing arsenic through the sediments piled up behind the dam and then through the Milltown aquifer at a rate of "2 to 20 pounds per day," which meant that it would take "several hundred years" to deplete the area of that one toxin, the first report continued to call the health risks "low." According to the RI, the only grave danger presented by the contamination was "substantial scour and transport of sediment," such as in a 100-year flood or dam failure.

Even though it would take hundreds of years for the problem to dissipate on its own, the report ranked possible centennial floods as "rare or unlikely events."[39]

Nevertheless, the six solutions remaining in the draft all left the dam in place. Four of the six, including "no action," were fixes that left the contaminated sediment in place as well. With these limited options on the table, the Milltown Technical Advisory Committee led the way in promoting public participation at upcoming meetings where EPA and ARCO representatives planned to provide extensive detail about the six options, set a time frame for choosing one, and take public comment. Therein Milltown continued to exemplify what was happening with Superfund across the nation. As the director of Montana's EPA office put it, "while greater and greater public input is occurring, a more streamlined, more cost-efficient approach is being mandated." In an open letter to concerned organizations and the public, John Wardell wrote that "Montana is recognized as a national model for public involvement in the Superfund process, which is a credit to many of you who have committed much time and talent." Wardell requested that people continue to participate in the process, especially as the Milltown FS was being finalized. He got what he asked for.[40]

Of all the reactions to the shortened list of six possible solutions, the state's surprised people the most. Since first filing a claim in 1983 for restoration money against ARCO, the state had spent $7 million on its case in twelve years. Representing Missoula, where he would become mayor the following year, Mike Kadas asked the 1995 state legislature to approve another $2.3 million to help fund, as he put it, "our one chance in history to bring this area back from a century of pollution." With near unanimity the legislature approved the expenditure. But what the recipient organization, the state's Natural Resource Damage Litigation Program (NRDLP), envisioned as restoration at Milltown was inadequate to most local observers. The state tentatively backed the remedial solution that called for leaving contaminated sediments in place behind the dam and managing them through "institutional controls." In the short run, that meant having the MPC maintain its dam, continue slow drawdowns, and control recreational use around the reservoir to limit people's risk of exposure to toxins. In the long run, it left the removal of heavy metals to nature. Financially, it relieved ARCO of much cleanup expense. For the state, it closed the door on actually restoring the Milltown stretch of the river to anything resembling a healthy, waste-free environment. And for Missoula, it spelled a persistent threat.[41]

The swiftest and strongest reaction to the state's support for an "in place" remedy came from downstream of Milltown. Perhaps it's not surprising that the area's largest community, located seven river miles below the Milltown Dam, opposed the idea of just leaving 6.6 million cubic yards of toxic sediment in place behind an old, leaky, partially wooden structure plugging up a river prone to flooding. The Missoula City-County Health Department reacted with a sixteen-point resolution. On February 16, 1995, the department resolved to "strongly oppose the recommendation of the Montana Natural Resource [Damage] Litigation Program that no action be taken to remove contamination sources or to prevent the spread of contamination from the Milltown Reservoir." Much of the resolution seemed to reiterate Superfund language, calling for "the most effective and permanent" solution and requesting "public comment" to help guide the decision. Individual residents wrote letters echoing the health department's position. Sent directly to the state's NRDLP, many of these letters cited human and environmental health concerns as well as recreation interests in their support for a more aggressive cleanup. They also saw corporate power at play in the decision. As one person put it, the deal looked like "a capitulation to ARCO and a perpetual danger to the people (and other creatures) downstream." Any kowtowing by the state to the company seemed unlikely given the legislature's recent passage of more funding for the NRDLP to continue its more-than-a-decade-long fight against ARCO, but people were angry.[42]

Most of the anger over the state's support of keeping dangerous sediments in place behind Milltown Dam was borne of fear. In every reaction to that decision, people wrote of the possibility of the dam collapsing, of a catastrophic flood, essentially of looming disaster. While hopes for a better environment inspired steady public participation throughout the Superfund process at Milltown, events or decisions that evoked a sense of fear resulted in the most public input. Even in a follow-up letter to the city's resolution, Peter Nielsen, the new environmental health supervisor for the health department, mixed scientifically based assertions with the scary ramifications of a dam failure or other event that could wash arsenic downstream.[43]

The Clark Fork Coalition and the Milltown Technical Advisory Committee went a step further in their opposition to the state's choice of remedies at Milltown. Both organizations questioned the entire winnowing process that eliminated nearly three-quarters of the original remedial alternatives. MTAC requested that the EPA revisit that process and pay particular attention to retaining "dam engineering" options, such as removal. Both groups argued that the state's preference for containing the wastes at Milltown was a poor fit with the overall Natural Resource Damage suit.

The coalition wrote that the suit's purpose, as dictated by Superfund law, was "to restore damaged natural resources to 'baseline conditions.'" The CFC charged the state with failing to define the basic conditions of the river that restoration should achieve, "such as the plant, fish, and wildlife communities they expect the basin to support." Unless the baseline condition of the rivers' confluence was a toxin-filled reservoir, then a keep-the-toxins-in-place solution seemed counter to the state's lawsuit, which had ballooned to $635 million by 1995. Some of that money was to go to improving the conditions of the river at Milltown, even if full restoration of the river to pre-European conditions was admittedly impossible. In addition, the CFC and MTAC asserted that the state's position ran contrary to Superfund's mandate to favor permanent remedies. In reaction to the state's claim that removing Milltown's toxic sediments ran the risk of flushing some of them downstream, CFC staff scientist Geoffrey Smith wrote that while "NRDP cites the potential for downstream injury as a reason not to remove the sediments, [it] fails to address the potential downstream impacts of leaving them in place." Again, this argument hinged on fear of a dam mishap during the countless lifetimes it would take for the toxins to "naturally" disperse. As MTAC put it, thousands of years is a long time to speculate that there won't be "catastrophic events involving floods and/or earthquakes."[44]

Just as the Missoula health department offered the earliest opposition to the state's support of a virtual "no action" plan at Milltown, it also put forth some of the most incisive critiques of that plan. Peter Nielsen confirmed that since the city designated its sole-source aquifer in 1988 and established a water-quality district in 1993, local businesses had spent considerable time and money complying with new regulations to maintain, if not restore, clean water. As he put it, "the community expectation has been clearly expressed." That expectation was that anyone affecting community water should be held liable. Nielsen contended that if the state settled for a "no action" alternative at Milltown it would seriously compromise the public's ability to sway the EPA toward any other remedial option. In other words, he charged the state with failing to uphold citizens' wishes. "When residents . . . supported the creation of the natural resource damage program . . . it was our clear understanding that the program would attempt to seek restoration of groundwater resources *above and beyond* the remedial actions taken through the EPA cleanup process." The NRDLP was supposed to pursue money to improve on Superfund cleanup, not hinder it.

Even with that, the state wasn't exactly rolling over for ARCO. In October 1995, the state upped its natural resource damage lawsuit by $78 million, to $713.3 million. ARCO vowed to fight the suit "to the very end." Although the company had the proverbial deep pockets that Superfund was meant to capitalize on, a year earlier ARCO had disclosed to the federal Securities and Exchange Commission that the $648 million it had set aside for environmental cleanups was about $1 billion shy of its newly estimated expenditures. If Montana's natural resource damage claim succeeded, the company would have to re-up its environmental expenditures by nearly another three-quarters of a billion dollars.[45]

Money certainly played a role in the debate about remedial alternatives at Milltown. The EPA recognized that the cost of removing sediments from behind the dam would be "quite high," as would the subsequent natural resource damage claims against ARCO. In the drafts of the RI and FS, ARCO penned the cost estimates for each remedial alternative and also ranked the alternatives based in part on their economic feasibility. That didn't mean that the company had free reign in figuring its possible future expenditures on cleanup. The EPA and the state DHES jointly set guidelines for ARCO's calculations and enforced revisions along the way. In responses to citizen and local organizations' concerns about the remedial alternatives, the agencies indicated no inclination to rule out possible solutions based solely, or even primarily, on cost. What the agencies did indicate was that local comments were changing the FS. One change that the EPA and state did not foresee was agency support for dam removal. The river would soon amend that.[46]

~ ~ ~

Back when the Federal Energy Regulatory Commission issued the Montana Power Company a six-year license extension, the agency indicated that if the Superfund process left toxic sediments in place behind the dam it would consider classifying the structure as "a high hazard dam." Throughout 1995 the release of FS drafts and the reactions to them put the issue of permanently leaving sediments behind Milltown Dam at the forefront of the Superfund process and the state's natural resource claim against ARCO. ARCO supported the "in place" remedy along with institutional controls, as did the state and EPA. Their reasoning hinged on mitigating the short-term impacts of trying to move thousands of tons of toxic sediment out of a reservoir without washing them downstream, combined with proposals to maintain the Milltown Dam indefinitely. Local opinion took a much longer view. Individuals and area organizations argued that short-term risks paled in comparison to the long-term possibility of failure to maintain a safe, stable dam. The issue of whether or not the EPA should require sediment removal was tied up with the NRDLP's suit against ARCO. The way that the EPA required the company to remediate would set the stage for the degree to which the state could restore the river. Leaving sediments in place left very little restorative possibility. All the study and observation of the river that the Superfund process and long state lawsuit had entailed helped people see a severely damaged watershed, even if few people donned masks and snorkels to emulate George Ochenski.[47]

Just before Christmas 1995, the EPA invited the public to offer another round of comments on the ongoing FS process. Earlier criticisms had pushed ARCO to rewrite the FS, and the EPA opened a window for public input on the new product from December 26 through March 15, 1996. ARCO released yet another draft of the FS in April 1996. The list of remedial alternatives had increased from six to eight. Two of those eight included removing the sediments behind the Milltown Dam. But the company had changed its assessment very little. In its final analysis, ARCO recommended the same option that it had before: leave the toxic sludge in place, use institutional controls to keep people out of harm's way, monitor the site indefinitely, and wait for nature to do the cleaning. But the April report was too late. Nature had already played its hand. During the comment period that preceded the release of the new draft FS, the Clark Fork and the Blackfoot River made the thought of permanently leaving hundreds of tons of toxic sediment behind the dam obsolete. It would take a while, as everything at Milltown seemed to do, for that obsolescence to become reality. But one day during the winter of 1996 the rivers flowing into the Milltown Reservoir once again forced the issue of their own cleanup.[48]

CHAPTER 5

Another River Runs through It

Paul and I fished a good many big rivers, but when one of us referred to "the big river" the other knew it was the Big Blackfoot. It isn't the biggest river we fished, but it is the most powerful, and per pound, so are its fish.
—Norman Maclean, *A River Runs through It*

The reason I came out here today to make this announcement is because I wanted to find the place—the one place in the West—that I thought told the most powerful story about the possibility for restoration.
—U.S. secretary of the interior Bruce Babbitt on the Blackfoot River (1998)

Robert Redford's film adaptation of *A River Runs through It* premiered in Montana in autumn 1992. Like Norman Maclean's 1976 novella, a semiautobiographical tale of a family growing up in Missoula, the film began with one of the book's two most quoted lines: "In our family, there was no clear line between religion and fly fishing." In the story that follows, the Blackfoot River is the scene of familial struggles, the religious zeal coursing through those struggles, and the love and art of fly-fishing. The Big Blackfoot River, as people called it in the novel's early-twentieth-century setting, helps stay those struggles and matches that zeal. The year after its publication, Maclean's short-story collection was nominated for a Pulitzer Prize, although the prize-granting board ultimately declined to bestow an award that year. Nevertheless, the story was legend in the fly-fishing community, especially in western Montana. Even before the movie's production or release, it was common for newspaper stories about "Norman's river" to begin with quips like "[the book] has kindled a national love affair with the Blackfoot River." But in 1992 it was the movie that catalyzed support for the story's real-life river and its legendary "high jumping" trout. The deep-rooted, as well as the newly gained, notoriety of the Blackfoot helped shape the fate of the Milltown Dam Superfund site in the late 1990s.[1]

U.S. secretary of the interior Bruce Babbitt paid a visit to the Blackfoot River in early June 1998. Plying the waters of the Clark Fork's biggest and most celebrated tributary for trout was as symbolic as it was genuine sport for Babbitt. He cast a dry fly on top of the current, which onlookers speculated had little chance of attracting a fish, but it made for better television footage than trolling under the surface with a nymph. Rather than trying to catch an actual trout, Babbitt was angling to allay fears that his agency's decision to apply the Endangered Species Act (ESA) to bull trout would not kill recreational use of western Montana's prized fishing grounds. In 1985 the U.S. Forest Service had deemed the bull trout a sensitive species, making it eligible for protection under the ESA. After studying the fish and its remaining habitat, a coalition of Montana-based environmental groups petitioned the U.S. Fish and Wildlife Service to list the species in 1992. Six years later, following much political and courtroom haggling, Babbitt disclosed the listing of bull trout as "threatened," meaning that his department would consider the fish for protection as "endangered" within a year.

Babbitt's trip to western Montana was as much about the Blackfoot River as it was about one of its signature species. The choice of backdrop for announcing the federal agency's decision to "conserve and restore" bull trout populations in the Columbia and Klamath River systems both acknowledged the fame of the river's recreational use and celebrated its revival. Babbitt told a crowd of government workers, area residents, and conservationists, "The reason I came out here today to make this announcement is because I wanted to find the place—the one place in the West—that I thought told the most powerful story about the possibility for restoration." He was referring specifically to the efforts of a coalition of people, called the Blackfoot Challenge, which had launched most of the restoration work and study of the river for the past decade. That Brad Pitt plied the waters in Redford's Hollywood version of the Blackfoot certainly lent renown to the river's story.

Listing the bull trout did more than aid the recovery of a single species in the Blackfoot. The 1998 listing included the warning that the federal government might add cutthroat trout to the list of threatened species within a year. It also drew attention to the Milltown Dam's effect on the survival of both fish in western Montana.[2]

In addition, the bull trout listing and Babbitt's visit to the Blackfoot were the seminal events that placed this Clark Fork tributary at the center of the Milltown story. During the 1990s the Blackfoot's health and future garnered as much local concern as those of the beleaguered Clark Fork. Coinciding with the listing of bull trout, Superfund-related studies of the Milltown Dam's effect on fish also drew a strong link between the structure and the Blackfoot. These events marked the turning point at which the remedy of leaving the dam and sediments in place at Milltown gave way to discussions of sediment and dam removal. In the bigger picture, what happened at Milltown because of the Blackfoot River and the attention it drew illustrated the flexibility within Superfund law, the permeability of Superfund sites, and the benefits of a long and sometimes sluggish Superfund process.

~ ~ ~

By 1996, the biggest worry about leaving thousands of tons of toxic sludge in Milltown Reservoir was the dam's stability. Everyone involved in finding a permanent remedy for the Milltown Superfund site knew that the river had battered the dam repeatedly in the nine decades since it was built. Locals worried about the inevitability of spring high water pounding an aging human-made structure. In the winter of 1996, as debate about Superfund remedies at Milltown heated up, the Clark Fork watershed altered that debate. It was not the Clark Fork proper, nor was it rushing water that did so. What focused new attention on the dam was the Blackfoot and ice.

The flood of ice that churned down the Blackfoot River toward Milltown in early February scoured the banks, drew crowds of onlookers, left one family homeless, and ultimately forced the EPA to reconsider its Superfund fix. The Blackfoot ice jam of 1996 also came at a time when the attention on the ecological health of that river was high. Some of the same people who stood along the river road to watch massive ice chunks grind, pop, and crunch past had been holding meetings to discuss the long-term problems of the Blackfoot River fishery for years. Soon they came to see the Milltown Dam as one of the problems, particularly for the newly listed "threatened" fish. The ice jam and the attention to the Blackfoot River fishery, especially bull trout, delayed and altered the Superfund cleanup remedy for the Milltown site. It also illustrated incongruities between the realities of a major Superfund cleanup project and national debates about the law. On the national level Superfund suffered criticism for its slow pace of implementation. Locally that pace allowed for a radical rethinking of cleanup. The 1996 ice jam, the high point of years of attention to the Blackfoot's fishery, and the federal listing of the bull trout coincided to make taking down the dam a legitimate option at Milltown.

Bull trout evolved in colder water than most of their salmonid relatives. They spawn and rear in the chilliest waters a stream can provide, near springs or mountain-fed tributaries. As adults they migrate through a wide and complex variety of riparian cover, from tangled nests of driftwood to deep pools and shady, undercut banks. As much as these fish need specific microenvironments, they need to be able to move between those environments. Biologists have found that some bull trout travel more than one hundred river miles between where they were born and return to spawn and where they spend most of their adult lives. The upper Clark Fork was ideal bull trout habitat for millennia. The icy waters of the Blackfoot and other mountain tributaries provided ideal spawning grounds. The main river's warmer waters offered many miles of adult habitat. Those places where colder waters joined warmer ones were particularly important. The Salish Indian name for the confluence of Rattlesnake Creek with the Clark Fork in Missoula meant "many bull trout." "More bull trout," or more commonly, "place of the big bull trout" was their name for the confluence of the Blackfoot with the Clark Fork. The Milltown Dam severed a migratory path along which bull trout had evolved. But its building had not erased the evolutionary memory of that path. Fish wanted to get past the dam.[3]

People wanted fish to get past the dam, too. Fish passage meant better fishing and a healthier river. The ESA required that fish get past the dam. The delays and seemingly sluggish progress at Milltown allowed for the intersection of these three things. Hence, the evolving solution for this Superfund site benefited from the elapse of time rather than floundered because of it, as was one of the complaints about Superfund nationally.

~ ~ ~

On February 9, 1996, Peter Nielsen drove toward Milltown from his office in the Missoula health department. Early that Friday morning he planned on collecting a few water samples from untested community wells. Heading east with the Clark Fork in view out his passenger-side window, Nielsen listened to a radio report about a major ice floe coming down the Blackfoot. When he reached the Milltown Dam, where the Clark Fork and Blackfoot converge, he saw "stunningly silver-gray water coming over the dam," which he thought "very odd." Out of curiosity he took a small sample of water from below the dam. Then he followed the Blackfoot east for about a mile above the reservoir to check out the reported ice. Along with "scores of drivers" and a film crew from the Montana Power Company, Nielsen watched what the *Missoulian* later described as "crusty tundras of ice and snow that made the Blackfoot River appear solid from bank to bank."[4]

But the river wasn't solid. Where Nielsen and others watched from the banks, the slow but steadily moving crush of frozen water had temporarily backed up at a sharp bend in the river. Later that evening, a section of the ice floe swept a century-old home and wood shop off their foundations, ruining both. As one local resident, Richard Frank, put it, "when it starts to move, there is nothing that can stop it." That's what worried the dam's owner. Although MPC spokesperson Cort Freeman later claimed that "we never expected the dam to fail at all," the company took immediate steps to minimize any damage that the ice might inflict. Engineers opened some of the dam's flow gates, quickly dropping the water level in the reservoir. The sudden drop in water level stalled and collapsed the miles of ice battering its way toward Milltown, propelled by a river flowing at about fourteen thousand cubic feet per second, or nearly ten times the Blackfoot's average. Colder temperatures in the following days further stabilized the moving mass and diminished the early spring runoff that was swelling the river. People's immediate attention turned to other statewide flooding, including the drowning of a Missoula-area student who fell into the Clark Fork and got trapped under the sheets of ice that lined its banks.[5]

In the days following Nielsen's trip up the Blackfoot to collect water samples, hundreds of National Guard troops along with resident volunteers mucked out ditches, stacked sandbags, and hauled valuables to contain or minimize the damage from widespread flooding around western Montana. The day or so of excitement near Milltown faded as quickly as the Blackfoot's diminishing flow undercut the ice jam, breaking it apart and floating it downstream in smaller chunks.[6]

By the first of March some of that excitement returned in the form of concern over effects stemming from how the Montana Power Company had tried to halt the giant jam. The day after the ice jams drew attention, the Missoula health department resampled water below the Milltown Dam and discovered what Nielsen described as "enormously high levels of copper" in the water. The local newspaper printed Nielsen's review of health department samples that contained forty times the state standard for waterborne copper, ten times that of zinc, and five times that of arsenic. Nielsen called the heavy metal numbers "serious" and "quite a bit worse than any data I've been able to find." The Clark Fork Coalition's staff scientist, Geoff Smith, saw the event as a concern for the "biological integrity of the Clark Fork and the safety of Missoula's aquifer." While both men praised the MPC for protecting the safety of the dam itself, they also recognized that preserving the dam with all the toxic metal behind it was a permanent problem, not a permanent solution. Nielsen broached the question that would soon reanimate the debate about what cleanup meant at Milltown when he asked, "Are we going to leave those sediments in place? Could this sort of thing happen again?"[7]

People began to weigh in on that question immediately. Montana Fish, Wildlife and Parks fisheries manager Dennis Workman reasoned that as long as the Milltown sediments were in place the river would periodically flush them downstream. But what happened in February 1996 was, he thought, "a major event." Workman estimated that the event killed "loads" of fish, even if the stuff Nielsen had wondered about as it poured over the dam made it impossible to see. Like people breathing heavily contaminated smoke that irritates their lungs, fish exposed to that quantity of sediment, let alone sediment laced with heavy metals, produce so much mucus in their gills that they literally drown in it. More fish would die throughout the year as they ate metal-poisoned insects, Workman predicted.[8]

Just as the ice jam had fascinated onlookers, its consequences angered them. According to Nielsen, "the EPA received more phone calls and letters about Milltown than any other Superfund site following the ice jam. The reaction was triggered by fish kills and it was trout fishermen." Summarizing local sentiment, the *Missoulian* printed an editorial entitled "Milltown Toxins Must Go." In answer to Nielsen's query about the ice jam being a recurrent problem, the editorial read, "It's easy to imagine river flows and ice conditions that might create far worse problems."

Although no one knew it at the time, the winter of 1996 would pale in comparison to that of the following year.[9]

The thought of a flood or dam failure was hardly new. But the reality of such an event changed the fate of the site more than the hypotheses. And the reality contrasted with previous estimates of what the dam could withstand. In 1996 the ice-choked Blackfoot flowing at 14,000 cfs had paled in comparison to the "probable maximum flood" levels that the MPC had estimated its dam could withstand. If 14,000 cfs caused minor structural damage and forced the company to dump reservoir water that resulted in alarmingly high levels of downstream toxicity, what would the 271,400 cfs that the MPC claimed the structure could tolerate do? Looking back at the 1908 flood, which peaked at 48,000 cfs and nearly wiped out what was then called Clark's Dam, a river running more than five times that volume was an absurdity. The last time that much water had flowed through the canyon where the dam stood was probably at the end of the last Ice Age, more than ten thousand years earlier. An independent contractor supposed that the dam could withstand a "100-year flood" of the more modest level of 50,000 cfs. More importantly, the contractor pointed out that even if the dam held, "these flows will cause scour and transport of the trapped sediments that are located throughout the Clark Fork River basin and also those that are trapped in Milltown Reservoir." The EPA had shared concerns of this nature, too. As early as 1992 the agency had compiled a literature review of "dam-breaks and sediment releases following dam-breaks" as part of the Milltown site risk assessment "to identify similar dams that have failed and document the consequences of failure, especially with regard to sediment released." The agency's most pertinent discovery was of a 1991 dam failure on Idaho's Middle Fork of the Boise River. There, a normal spring flood had washed out a dam of similar age and construction as Milltown's, along with a quarter-million cubic yards of sediment contaminated with heavy metals from a century of upstream mining. The parallels with Milltown were uncanny, except that the Kirby Dam was much smaller and its sediment load a fraction of what Milltown impounded. And no Missoula sat immediately downstream of the dam.[10]

The events of 1996 hardly compared to a complete dam failure. But they made many people realize that catastrophe was a real possibility. By April, the Missoula City-County Health Department had released the official results of its Clark Fork water sampling from the previous two months. The report confirmed Nielsen's earlier assessments. It also gave the public and involved parties some more tangible ways of seeing the events of February than the rather abstract counts of heavy metal molecules per million. The ice jam measured "10 foot high, 40 feet wide and five miles long." Its movement down the Blackfoot at about "8–10 miles per hour" left approximately half a foot of sediment in places along the riverbank below the dam. Along with charts showing the specifics of the contaminants the ice floe had flushed downstream, the report made it clear that the event had essentially extended the Milltown Superfund site downstream of the dam. Calls for a different solution than leaving sediments behind Milltown Dam accelerated. But those calls emphasized the need for patience.[11]

Nationally the EPA received plenty of criticism in its first decade and a half for implementing Superfund projects too slowly. But a close look at Milltown revealed that taking more time, not less, resulted in a more thorough cleanup. And while local people may have urged the EPA to implement some solutions as quickly as possible, they also understood the need for patience in achieving the best long-term solution. The local chapter of Trout Unlimited exemplified as much when members penned a letter to EPA director Carol Browner stating, "We ask you to delay your decision and to, please, select another alternative than leaving the sediments in place untreated. Milltown has been on the National Priority List since 1982 so there is no need to rush now."[12]

Within a month of the ice event, the Missoula City Council teamed up with Peter Nielsen and the health department to write a resolution requesting a more permanent fix at Milltown than leaving sediments in place and resolving that the EPA should delay its decision. When the City Council passed the resolution on March 21, 1996, it included the argument that local businesses and municipal facilities had to abide by water-quality standards, so the Superfund remedy should as well. Predictable violations of those standards, such as happened in February because of sediment scouring, were an unacceptable solution for Milltown. So, too, the group resolved that the EPA should not cap the price tag of a solution. The resolution ended with a request for a delay in the process. The EPA fielded similar letters from Milltown and Missoula citizens, including one from a high school science class. Another letter, consigned by fifteen people, ended with the request that the agency avoid hurrying and instead "continue" with "careful planning to get the best result."[13]

The EPA responded. Agency director John Wardell let the local Trout Unlimited group know that his agency had "received several comments . . .

requesting delay of the feasibility study and decision for the Milltown Reservoir Site. In light of this public comment and the February 'scouring episode,' we are taking this request very seriously." Peter Nielsen recalled that in a private conversation with the city-county commissioners, Wardell had said, "We've decided to re-examine our thinking." Knowing that the regional director often resorted to understatement, Nielsen recognized that this meant that sediment removal and possibly dam removal were viable options. Wardell's comment appeared in the local newspaper on May Day 1996, although the paper reported him as saying that the EPA was unlikely "to recommend removal of the 6.6 million cubic yards of metals-polluted sediment in the reservoir." The estimated $134 million price tag was prohibitive. What the EPA did do was delay the feasibility study, just as so many people had urged. That delay aimed to address how the Milltown site was affecting the river below the dam and what solving that added problem entailed.[14]

~ ~ ~

Support for delaying a Superfund cleanup plan, especially one that was part of the largest such complex in the country, ran counter to the national dialogue on toxic site cleanup at the turn of the twenty-first century. One thing that both sides of a very partisan national dialogue agreed on was advocating faster cleanups. By a few key indicators, such as number of toxic sites identified, remedial actions taken and completed, and site expenditures, the Clinton years were the high point for Superfund. During most of the 1990s the EPA added sites to the NPL more rapidly than during any other period. So, too, the agency remediated and removed sites from that list at a faster rate than before. And the cost of site cleanups dipped. The EPA gave credit for these signs of success to its emphasis on prioritizing and using risk assessments to drive listing and cleanup decisions. National news stories and analysis touted faster and cheaper cleanups as an improvement in Superfund implementation.

One reason for such success was the Clinton administration's support of EPA funding. Threatening to veto any general budget bill that cut the agency's funding, Clinton forced Congress to leave the EPA pocketbook intact during heated negotiations over the federal government's spending throughout the 1996 legislative session, when Superfund was up for renewal. According to the *New York Times*, the executive victory on environmental issues was one of the few things the White House could "crow about." Regardless, congressional Republicans, such as House Speaker

Newt Gingrich, continued their barrage of criticism, charging that "Super-fund . . . is one of the most flawed programs ever designed in the name of environmental protection." Using selective statistics, Gingrich added that "over 16 years we have spent $15 billion and cleaned up fewer than 100 sites out of 1,600." The Speaker failed to account for the tens of thousands of sites the EPA had cleaned up without ever listing them on the NPL, or to acknowledge that most of the 1,600 or so on his list were deep into the cleanup process. From another perspective, "twice as many cleanups have been completed in the five years of the Clinton Administration than in the previous 12 years," reported the *Times* in 1998. And as Milltown illus-trated, in some places people advocated patience over haste when dealing with their waste. Yet what Gingrich and other critics most conflated was the issue of Superfund expenditures. He and other critics implied that the EPA was wasting public funds, when in fact nearly every cent of the Superfund budget came from special taxes on polluters or liability suits against PRPs. The "polluter pays" provision worked. And despite the sharp contrast of opinions on how to change Superfund, Congress renewed the program with virtually no amendments through the 1990s, mostly by sim-ply reauthorizing its funding scheme. And against stiff opposition, Clinton oversaw a continuation of Superfund's primary principle of the polluter paying for toxic cleanups.[15]

Republican lawmakers tried and failed to undermine the tradition of corporate responsibility in Superfund in a variety of ways. A 1997 defense bill included a provision that "would have exempted W. R. Grace & Com-pany, the former operator of a Federal Superfund site, from further li-ability for cleanup costs" at a site where the giant international chemical manufacturing company had paid only $800,000 of an expected $120 mil-lion bill. In an example that reverberated throughout Montana, Republi-cans in Congress tried to exempt the Clark Fork from a provision of their Superfund renewal bill that limited states' abilities to sue PRPs for natural resource damage claims. They did so because Montana's $765 million suit against ARCO was still pending and Republicans needed the support of Montana senator Max Baucus, the ranking Democrat on the Environ-mental Committee that was vetting the bill. According to the *New York Times*, "if Mr. Baucus was impressed by the gambit, he did not show it at the hearing." The bill failed, leaving all states with the opportunity to press for restoration funding from responsible parties. Through the end of the Clinton administration, the president, congressional Democrats, and EPA administrators combined to keep Superfund intact. And the EPA fostered a more efficient program by most measurements through administration choices. One such choice continued to be integrating public opinion into each stage of the cleanup process. That was certainly true when it came to how the EPA responded to the aftermath of the 1996 ice-scouring episode at Milltown.[16]

~ ~ ~

The EPA's first effort to find a new solution to the downstream contamination caused by the partially frozen Blackfoot was to divide the Milltown site into two cleanup decisions. One would address the contaminated aquifer; the other, the problems of downstream contamination. By doing so, the agency hoped to salvage the first feasibility study with its option of leaving sediments in place as the solution to the aquifer problem. Then the agency would consider a different solution to fix the troubles below the dam, such as skimming the most mobile sediments off the top of the reservoir's load of 6.6 million cubic yards. The agency began by requesting public opinion.[17]

With one notable exception, people opposed the idea of breaking Milltown's cleanup into two parts. The Montana Power Company supported the proposed split. A month later, in October 1996, the EPA sent ARCO comments on the draft feasibility study due for official release in December. The comments therein indicated that the agency assumed its Milltown splitting proposal would gain public approval. It was wrong. The CFC wrote in mid-November opposing the plan. Restating the request for a delay, the organization's director, Meg Nelson, emphasized their belief that the worry about downstream contamination was linked to the entire Milltown site, not just the top layer of easily mobile sediments. The group favored a single, thorough solution that would address all the risks posed by the dam and its sediments. The Missoula Board of County Commissioners seconded that opposition and the views substantiating it. MTAC weighted in similarly.

On December 23 the EPA announced that such comments had swayed it against separating the site into two parts. Instead, the agency would delay the whole process while it, the state, the PRP, and other concerned organizations pursued a new "focused Feasibility Study to address the release of sediments from the Milltown Reservoir" as well as from the groundwater. The new timeline predicted that the focused study would culminate "between January 1997 and early 1998."[18]

~ ~ ~

On May 9 and 10, 1996, 650 sixth graders from Missoula and the Milltown area attended class on the banks of the Clark Fork below the Milltown Dam. They spent their two days of open-air study learning about water. Activities and workshops helped them visualize the movement of individual water molecules through the cycle of evaporation, transpiration, condensation, and precipitation. They played games in which each student took on the role of a major aquatic component in the local watershed. Some kids, acting like clean water, helped guide their classmates, who mimicked native fish, through the perils of those who were approximating rocks, log jams, toxic chemicals, and a dam. The watershed became an obstacle course of friends and foes. Amy Michels recalled playing an insect that "had to die because of the pollution or chemical we were killed by." A special section of the *Missoulian* devoted to area students featured Hank Schewe's acrostic poem about one of western Montana's native fish. Reflecting some of the things he'd learned, Schewe wrote:

> Becoming extinct
> Underwater life
> Lives by eating bugs
> Lives a life of danger
> Trout family
> Really hard to tell the difference
> Over dams
> Under rocks
> Trouble.

The fact that hundreds of grade-schoolers were learning about the perils of life in local rivers was more than the product of an isolated day out of the classroom.[19]

In the 1990s the Blackfoot River got a great deal of attention. And as with so many environmental efforts and issues, a single species came to symbolize a place's problems and people's hopes for its future. In the case of the Blackfoot, that species was the bull trout. A host of local, state, and federal environmental agencies, businesses, and organizations sponsored the Watershed Festival that gathered Michels, Schewe, and their peers to the Clark Fork. With sponsors like the U.S. Forest Service; Montana Fish, Wildlife and Parks; the local natural history center and city health department; Montana Power; and the local mall, the event indicated the range of interests at stake. Its emphasis on children revealed community-wide concern and the degree to which the watershed's problems were becoming common knowledge. And the content of the festival left no doubt that one of the key things to know about the local river system was that it had problems alongside great potential. The most frequently noted symptom of those problems was a lack of fish. Summarizing this focus on fish, Montana FWP wrote to John Wardell in May 1996 that "the demand for wild trout fishing is growing but our ability to increase opportunities is limited." Potentially, the letter continued, "one of the ways we could accommodate more demand is to improve fish populations in the Clark Fork where they are depressed." Before and after 1996, FWP along with other organizations had been trying to make those improvements, especially by attending to the nationally famous trout stream, the Blackfoot River.[20]

The efforts to improve the Blackfoot River's fishery were connected to the Milltown Superfund site. By early 1997 the EPA had a list of the initiatives it had taken to address the "1996 ice scouring event which was the incident that brought the potential severity of such events to light." Almost all of those efforts aimed to curtail the death of fish downstream from the dam. That work resonated with what had been happening on the Blackfoot for years and would soon involve the Milltown Dam.

What had been occurring on the Blackfoot became a clear example of the principle of unforeseen consequences in Superfund sites. In the early 1990s the decline of the Blackfoot's fishery garnered growing attention, which eventually led to studies of how the Milltown Dam impacted that river and not just the Clark Fork. The historical and cultural importance of the Blackfoot inspired far more people to get involved with the Superfund decision at Milltown than was the case when the Clark Fork was the dominant focus. The other river that ran through the Milltown site made a big difference.

~ ~ ~

Historically, studies of the Clark Fork showed that the biggest reason that a modicum of fish managed to survive in the river was its tributaries. While mature fish, whether native or not, had very little tolerance for the rates and kinds of pollution in the main river, young fry had even less. So most of the viable spawning habitat that regenerated the Clark Fork fishery was in the colder, faster-moving streams and rivers that fed it. Above Missoula, the Blackfoot was the largest of these. In fact, historical averages show that the Blackfoot, at 1,606 cfs, carried more water than the Clark Fork's 1,369 cfs. For most of the year, including at flood stages, the Blackfoot more than doubled the size of the Clark Fork at the same point that it gave up its name, somewhere beneath the Milltown Reservoir. By burying the confluence of both rivers, the dam and reservoir were degrading the Blackfoot's ability to regenerate its fishery. Or as Hank Schewe put it, "over dams . . . trouble."[21]

People began considering how to mitigate the dam's effects on migrating fish during the Montana Power Company's efforts to relicense its dam in the 1980s. Because of pressure by groups like the CFC over fish passage, the license extension that the MPC finally attained included requirements that the company study and fund fish mitigation.[22]

At the same time, the Blackfoot River came under organized scrutiny. In 1988 a new chapter of Trout Unlimited formed on the Blackfoot River and launched "a five-year, $220,000 study." The chapter, formed by concerned fishermen, raised money to help fund the work of state FWP fisheries biologists hoping to "pinpoint why fishermen have been reporting declining catches of trout in the Big Blackfoot the past few years." FWP biologist Don Peters speculated that the river's recent lack of fish had multiple causes, including elevated levels of silt from soil erosion caused by excessive logging, heavy metals from historic mining in and around the Blackfoot's headwaters, and overfishing. Within a year, Peters's speculations proved fairly accurate. The continual industrial uses of the river basin were damaging water quality; what remained, people loved to death.[23]

The first solutions that FWP biologists suggested were to regulate fishing and fix some of the Blackfoot's tributaries. Don Peters proposed making cutthroat and bull trout catch-and-release to help keep breeding fish in the river. According to studies of what Peters called "the crash"—the decline of the Blackfoot fishery after the 1970s—the two native fish, cutthroats and bull trout, were probably at all-time population lows. Averaging 1.8 cutthroats and 0.2 bull trout per thousand feet, the river wasn't providing many opportunities to anglers anyway. The new policy would require the very deft or lucky fishermen who actually caught a native trout to throw it back. The strategy aimed to keep breeding fish in the river. Along with more fish, the river needed mending.[24]

In the next few years studies revealed very few surprises about what was harming the Blackfoot fishery. Steps to fix it began to succeed. Logging plagued the lower river. Since before William Clark had funded the Mill-town Dam, whole logs had choked the Blackfoot's lower stretches. Spring high water floated thousands of winter-cut trees toward the lumber mills at the Blackfoot's confluence with the Clark Fork, ravaging the river's bed and banks for miles and devastating spawning runs. Sediment from cutover soil rushed into the river along with the seasonal thawing of moun-tain snowpack, clogging the river's gravel beds where trout laid their eggs and where the insects on which they fed spent most of their lives. Modern logging substituted roads and trucks for the river. During the 1970s and '80s the U.S. Forest Service built 340,000 miles of road, mostly in western national forests. At eight times the length of all interstate highways, these roads made the Forest Service the country's most prolific road builder. It also enabled modern corporate logging on public lands. Through the 1980s logging roads in the Blackfoot gave industrial logging companies access to tens of thousands of acres, which they commonly clear-cut, leaving denuded hillsides and road cuts behind, both of which acceler-ated sedimentation into the river below. In *Last Stand*, a book about il-legal clear-cutting practices in the Blackfoot Valley, *Missoulian* journalist Richard Manning described how "silt gathers in the spaces between rocks where trout deposit eggs, smothering both the eggs and young trout. In a few years, once vibrant trout streams become sterile ditches, broken by the work of carrying topsoil away." He based that description on what he, scores of trout fishermen, and a growing contingent of fisheries biologists had observed in the Blackfoot River.[25]

The upper river suffered from its mining legacy. In particular, an old dam that held back heavy metal–contaminated water and sediment burst in 1975. By 1988 studies showed that native fish populations were still "20 times less than in a 1973 study." What few fish remained in the whole river bore the brunt of heavy fishing.[26]

By 1991, the Blackfoot had improved. As a story in the local outdoors section of the newspaper speculated, all the press about how poor the river fishing was might have helped keep anglers away. Fewer fishermen, along-side Don Peters's creel limits on native fish, combined to help more fish live and reproduce. With a few high-water years to wash out some of the sediment accumulation in the main channel, Peters predicted, the river might rival Rock Creek, a nationally renowned trout stream that feeds the Clark Fork above its confluence with the Blackfoot. "It's better big-fish habitat. The bug life is there, the deep holes, I think it's going to produce big fish consistently," he said. The Blackfoot was also about to gain a big audience, most of whom cheered its recovery.[27]

In *A River Runs through It* the Blackfoot River was too compromised to play itself. The first showing of Redford's movie was in Bozeman, Montana, a three-hour drive southeast and over the Continental Divide from Missoula. The choice to screen the movie outside Missoula followed from the earlier choice to exclude the real Blackfoot from the making of the film. When producers toured the river that ran through Maclean's tale they deemed it too degraded to star on the big screen. So they shot the movie's river scenes on the Gallatin River south of Bozeman on the headwaters of the Missouri River rather than those of the Columbia. While that irked many people in western Montana when it came time to debut the movie, they seemed to understand. The Blackfoot was not pretty enough for Hollywood. But they also hoped that its portrayal by a worthy double might help repair it. A month after the premiere, the Big Blackfoot chapter of Trout Unlimited (TU), the CFC, and a local fly-fishing shop sponsored a benefit screening of *A River Runs through It* in Missoula to raise money for the river restoration projects they were helping organize and oversee. Nationally, the movie paid immediate dividends to the real Blackfoot as well. By the spring of 1993, the river's TU chapter had received nearly $200,000 from donations collected by Orvis, an outdoor gear company specializing in fly-fishing equipment. The company had advertised the Blackfoot's problems and needs in national outdoor magazines and offered donors of more than $150 "a special edition copy of Norman Maclean's book." Anyone giving double that amount received a copy of the book *The Making of A River Runs through It* signed by Robert Redford. The Orvis money helped TU toward its goal of $400,000 for the Blackfoot projects, which would trigger the National Fish and Wildlife Foundation to contribute another $200,000.[28]

The same people were involved in deciding how money should be spent to deal with fish at Milltown Dam. In late October 1992, the Montana Power Company released a draft of its "Milltown Fisheries Protection, Mitigation, and Enhancement Plan," which its license extension had required. The Westslope chapter of Montana's TU, the CFC, and the state's FWP all consulted with the MPC on the plan, along with numerous other federal, state, and local agencies. The draft plan estimated that the dam blocked passage of spawning fish for "56.7 miles of the Clark Fork River . . . and 13.5 miles of the Blackfoot." The dam impeded upstream migration of fish spawning either in the spring (rainbow and cutthroat trout) or the fall (bull and brown trout). But those estimates lacked solid data on actual fish; they were guesses based on comparing the rivers flowing into Milltown with other rivers in Montana. The proposed solution of slowing dam drawdowns, funding hatchery fish to replace the estimated spawning losses, transporting some fish over the dam, restoring tributaries on both rivers to improve spawning habitat, and of course more study, would all cost the MPC $65,000 per year.[29]

Within a year the price tag had risen and the emphasis had shifted toward more research. For an additional $50,000 the company offered to fund a fish ladder that would allow biologists to trap migratory fish and, in particular, "separate bull trout from other species and tag them to track their migratory behavior." Dennis Workman, a state FWP fisheries biologist who had worked closely with Don Peters for years studying fish in the Clark Fork watershed, praised the MPC's efforts, saying, "You can't ask for much more." Although most environmental groups ultimately opposed the idea of enhancing the rivers with hatchery fish, as well as establishing a fixed financial obligation for the MPC in perpetuity, the rest of the plan continued to meet with approval. As the CFC put it in a letter to the MPC, "emphasis in the plan should be placed on natural habitat improvements not artificial quick-fix approaches." The plan was one more way that people began to connect the dam with the Blackfoot, a connection that would later change the course of the Superfund solution at Milltown. In spite of estimates that the dam was preventing "from 75 trout to 750 trout" per river mile from pursuing their instinctual migrations, Don Sprague, manager of the MPC's environmental department, said that dam removal was "not an option" because of the sediment it harbored. At least not yet it wasn't.[30]

Although the MPC opposed dam removal, in spite of that solution's inclusion in the ongoing FS for Milltown, the company did court broad comment. In a letter announcing the release of its final plan, the MPC noted that it was required to include comments from only a limited number of state agencies, but it had voluntarily solicited and incorporated "full participation," which allowed it to "reach substantial agreement on many issues." One of the newest participants was the Confederated Salish and Kootenai Tribes (CSKT).[31]

The tribes' participation in the Superfund process at Milltown would grow to include restoration work on tribal lands and joining the state's suit against ARCO, but it began with concern over fish. Along with *A River Runs through It*, the most commonly known cultural reference locals made about the Blackfoot came from tribal history. Long before the Milltown Dam drowned the confluence of the Blackfoot and Clark Fork to produce electricity, Salish-speaking people used the area during their seasonal ventures from the valleys radiating out from Missoula to the bison hunting grounds of eastern Montana. As they traveled up the Clark Fork, bison hunting parties crossed the river's junction with present-day Rattlesnake Creek, which they called *Nɫʔay(cčstm)*, or "place of the small bull trout." A few miles upstream from there they left the Clark Fork and began following the Blackfoot at *Nʔaycčstm*. This "place of the big bull trout" offered a fine camping spot where the Salish stopped to fish the waters of the confluence before their journey over the Continental Divide.[32]

Although the confluence of the Blackfoot and Clark Fork was a unique place in Salish tribal history, it was less unique as a general environment for bull trout. In spite of their common name, bull trout are not trout. They are char, a subgroup of salmon. Their taxonomic misrepresentation as trout stems from a long history of refining the biological boundaries of different fish species. True arctic char, like bull trout, lack the teeth in the roof of their mouths that true trout possess. Through that long process of distinguishing fish species and naming them, bull trout acquired the Latin name *Salvelinus confluentus* in the mid-nineteenth century. The English translation of the Linnaean genus and species appellation is roughly "brook trout" and "flowing together." The former refers to the bull trout's early lumping with trout, and the latter to its preference for spending most of its mature life in larger waters, especially the confluence of rivers that provide access to cold spawning grounds and warmer wintering ones. The fish's Latin name, much like the Salish place names, relates bull trout to the meeting of streams.[33]

The involvement of the Confederated Salish and Kootenai Tribes in the Milltown Superfund process hinged on the tribes' historic use of places like the rivers' confluence. The 1855 Treaty of Hellgate had pushed a confederation of Salish, Kootenai, and Pend d'Oreille tribes onto a reservation north of the Clark Fork in Montana. The founding of the Flathead Reservation also preserved the tribes' access to traditional hunting and gathering grounds in western Montana, including the "place of the big bull trout." So in 1993, when the CSKT commented on the MPC's plan to protect, mitigate, and enhance fish at the dam site, they were asserting "legal interests to all aquatic resources associated with the Clark Fork River." In a letter to the MPC, tribal council chairman Michael T. Pablo accepted most of the company's plans but argued that the "commitment must be for a long term," especially in regard to bull trout. The long history the tribes had with the Milltown area and what was becoming its signature species, the bull trout, would soon dovetail with the most thorough study of the dam's effects on fish and the listing of the bull trout as an endangered species by the federal government. All of these shaped Superfund decisions at Milltown. This was one of those times when the letter of the law could not have predicted what was going to happen. Rather, the process unfolded because local conditions triggered or emphasized particular provisions within the Superfund law, including its provisions demanding public input, according tribal sovereignty, and accounting for other environmental laws, such as the Endangered Species Act.[34]

~ ~ ~

Usually, the Clark Fork fish that people put in a cooler were headed for the dinner table or freezer. In the spring of 1995, coolers full of fish started helping Clark Fork fish survive rather than become a meal. With MPC money, Montana FWP biologists rigged the company's dam with a trap meant to catch native fish on their upstream spawning runs. As part of the MPC mitigation plan, biologists used the trap to capture trout and transport them around the dam in coolers because there were so few spawning-quality tributaries below the structure and so many good ones above it. Biologists doing the work termed bull trout "at risk of extinction throughout their historic range, and westslope cutthroat . . . 'a heartbeat away.'" The *Missoulian* article that first covered the story mentioned the importance of the rivers' junction in Salish history, as well as the electro-tagging of captured fish that would allow researchers to study their future.[35]

Fish making it around Milltown Dam with a little help from biologists were meant to help the ongoing efforts to revive the Blackfoot as much as be a part of the Clark Fork Superfund process. A 1995 retrospective on those efforts celebrated the work of Blackfoot Valley residents in raising money and awareness, which had amounted to the restoration of hundreds of miles of the river's tributaries, wetlands, and main channel. That work began in the late 1970s with a loosely knit group of valley landowners who came together with ideas about how to enhance the Blackfoot watershed's natural resources as well as its rural character. By 1993 the partnership named itself the Blackfoot Challenge. Remaining locally organized and gaining national renown, the Challenge has managed to foster the reintroduction of wolves and grizzly bears into the basin, while also maintaining its ranching heritage and tailoring logging practices toward better forest health. On the river itself, its members worked with government agencies and environmental groups to create log and debris dams meant to replicate what beavers once did to establish a diverse ecosystem that favored trout reproduction. Repairing tributaries from overgrazing, keeping cows and logging out of the most sensitive riparian areas, and making irrigation more efficient so more water stayed in the river were all examples of their efforts. Most of that work aimed to undo what a century or more of mining and its symbiotic industries of logging and agriculture had done to the river. But garnering support for the Blackfoot's restoration also made people more willing to accept restrictive fishing regulations that helped reverse the overharvesting that had resulted from the waterway's popularity. In an ironic but much needed twist, Redford's movie and the resurgence in popularity of Maclean's book encouraged support for the Blackfoot's health, which decades of overzealous fly-fishing had diminished.[36]

But celebrations of the Blackfoot's future were perhaps premature. While mining had caused most of the damage to the river, mining in the river valley was not just history. In 1996, just after ice threatened to take out Milltown Dam, the national environmental group American Rivers placed the Blackfoot River on its annual list of the nation's most threatened rivers. Four years earlier the same group had listed the Blackfoot as one of the nation's most endangered rivers. The leap from being one of many endangered rivers, which meant that it had an "imminent risk of destruction," to being the most threatened, signifying its position at the "brink" of demise, was the result of mining. Just as things were looking better for the Blackfoot, a major gold-mining proposal threatened to turn back those efforts. Plans for an open-pit mine above the Blackfoot's headwaters near Lincoln, Montana (which had acquired its own sort of fame when the FBI arrested Ted Kaczynski, a.k.a. the Unabomber, in his handcrafted cabin outside Lincoln in early April 1996), included the practice of pouring cyanide over heaps of raw ore to extract the precious metal. That threat got the Blackfoot listed above such major watersheds as the Florida Everglades, the mighty and maligned Mississippi River, and the urbanized Hackensack where it runs through New Jersey and New York.

National prominence for the Blackfoot's contemporary challenge came from local effort. It was the Montana Environmental Information Center that nominated the Blackfoot for the yearly American Rivers "award." And it became a mostly local effort to protect the river against the mine proposal. The state's governor at the time, Marc Racicot, had put together a bull trout restoration team two years earlier to aid in the revival of the Blackfoot. He did so because of local encouragement in the face of the unwillingness of the U.S. Fish and Wildlife Service (USFWS) to create a federal plan for the species. Montana's bull trout restoration plan was one of the first and most well-regarded state-sponsored efforts on behalf of the fish. It helped push the federal government to apply the Endangered Species Act to the bull trout. As Glen Marx, the director of the state team, put it, "we want to prove we can manage bull trout. . . . The bull trout affects our economy, our quality of life, our recreation. Why shouldn't we take the lead in its recovery?" Marx hit on the trilogy of motives that infused most environmental organizations by the 1990s and pointed in the direction of local or watershed management that many of those same groups favored. But this wasn't the fruits of a states' rights movement, such as the Sagebrush Rebellion of the Reagan years. Local groups and state agencies consistently looked for support within a federal framework. Racicot's team ranged from the Bonneville Power Administration and major timber interests to the CSKT, numerous state agencies, and environmental organizations. In addition to helping restore bull trout, the team aimed to push the USFWS to consider the species for ESA designation.

With the Blackfoot River and bull trout restoration at stake, Racicot joined a chorus of concern over the proposed Seven-Up Pete mine. The CFC and others denounced the mine's proposal in an editorial to the Missoula newspaper that urged people to go see the proposed mine site for themselves. The group asked the public to imagine a "600-foot stack of waste rock, 895 acres of cyanide heap-leach pads, [and] an almost square-mile wide pit" marring the recuperating Blackfoot's upper reaches. Of most prominent concern was the river's rebounding fishery.[37]

The first major study of the mine proposal determined that the Blackfoot's headwaters were its richest spawning grounds and the ones most imperiled by the proposed mine. In general, the river's richness in terms of fish increased with distance from the Milltown Dam. So, even as people fought against an imminent threat to the river, they recognized its biggest ongoing problem.[38]

In 1997 American Rivers relisted the Blackfoot as one of the country's most threatened rivers, with the Missouri River replacing it in the top spot.

Even so, the Blackfoot's cause was about to gain a very important ally in the form of more federal environmental law. Just as with the Superfund process that was beginning to tie the Blackfoot more securely to the fate of Milltown and the Clark Fork, the designation of the bull trout under the ESA hinged on local efforts. By century's end, ongoing local efforts along with damning environmental impact statements would stop the Seven-Up Pete mine proposal. That was no small victory for advocates of a clean Blackfoot. Phelps Dodge owned the Blackfoot mining venture and poured $2 million into a campaign that killed a 1996 citizens' ballot initiative in Montana that would have made cyanide extraction of gold illegal. In his book on the history of the Blackfoot and that particular initiative, journalist Richard Manning quoted the state's Trout Unlimited director, Bruce Farling, as saying, "We could have come up with an initiative that said 'Don't rape grandmothers.' If those guys opposed it with $2 million, it would have lost." But losing the fight to ban cyanide heap-leach mining outright failed to kill the more specific effort to disallow the actual mine. That was largely because of what Manning referred to as the maturity of local environmental groups like the CFC, groups that focused on whole regions or watersheds rather than on single issues; that grew rather than wilted through transitions in leadership and staff; and that worked, by and large, with stable government agencies rather than at odds with them.[39]

Beginning in 1997, local environmental and governmental efforts coincided to support the listing of the bull trout on the endangered species list. That effort, along with the ice jam of 1996, made the idea of maintaining the Milltown Dam as part of the Superfund remedy substantially less tenable. At a summer public comment session the USFWS held in Missoula on the issue of the bull trout's ESA listing, only two of eighteen speakers opposed the measure. Both were timber industry representatives. Those in favor were, according to the *Missoulian*, "conservation organizations." These and other Montana environmental advocates saw listing the bull trout as the means to protecting an entire ecosystem. A fisheries biologist for the USFWS shared that assessment. In fact, the use of endangered species as a tool for preserving ecosystems had become the ESA's most powerful and unexpected ramification. Indicator species such as the bull trout stood in for the well-being of their native environments, like the Blackfoot watershed.[40]

Studies detailing the importance of the Blackfoot for bull trout spawning began to pile up. A lot of that work was thanks to Montana FWP fisheries biologist Don Peters, who vigorously opposed the plan to mine gold in the Blackfoot Valley and leach it out of ore with cyanide. His best weapon in fighting the mine was getting his agency to produce "a thorough and ongoing analysis of fish populations in the upper Blackfoot and its . . . tributaries." He reasoned, according to the *Missoulian*, that "we've put half-a-million dollars a year into projects designed to improve habitat . . . in the Blackfoot. . . . And, in fact, it [the mine] has the great potential to harm a lot of our good work." By April 1998, the Blackfoot was on the American Rivers list of most endangered rivers once again, with upstream mining the major worry. But by June, the presence of a single fisherman casting a dry fly into the river about halfway between the proposed mine location and the Milltown Dam would greatly lessen the gold mine's chances. Bruce Babbitt's announcement that the federal government had, after nearly a decade and a half of indecision, listed the bull trout under the ESA helped do just that.[41]

But for many people, a fish named "Pollywog" did more than anything to bring attention to the connection between the Blackfoot River and the Milltown Dam and to erode the idea of leaving the dam in place. Pollywog was a product of Superfund-related research at and on the dam. The story of Pollywog further demonstrates the importance in Superfund implementation of corporate funds, science, individual efforts, and public perception.

~ ~ ~

Like Peter Nielsen, David Schmetterling came to graduate school at the University of Montana for its location as much as for its academic offerings. Both men loved the school's proximity to rivers. Schmetterling grew up in Maryland, where he spent as much of his youth as possible playing outdoors and "catching things," especially fish. He used field guides and encyclopedias to make a list of all the fish he wanted to catch in their native waters as a way "to explore different places and the country." That list included cutthroat and bull trout. In the early 1990s he moved to Montana in pursuit of both a master's degree in fisheries biology at UM and the state's native fish.

After graduating and catching his share of fish out of the local rivers, Schmetterling began working for the state's Fish, Wildlife and Parks in 1995 out of the agency's Missoula office. As he remembers, "apropos to Milltown, it was a really interesting time." His initial work was "with native species in the Blackfoot and Clark Fork Rivers and understanding ways that anthropogenic influences across the landscape were influencing fish like cutthroat and bull trout." In the midst of that work, "Milltown Dam kept coming up even though we weren't looking at it," he said. There was very little solid information on how the dam affected fish. Schmetterling surmised that the lack of information was because "it just seemed like a hopeless situation." His work soon changed that perception.

In 1998 Schmetterling started a research project on the dam. He asked what he thought were a few basic questions: Was the dam still impeding fish migration almost a century after it was built? If so, what fish was it affecting? Was fish passage necessary? He asked these questions just as the MPC was finalizing its fish mitigation plan. The plan concluded that the dam altered fish life cycles for a fifteen-mile stretch of river at most. Schmetterling's findings forced a radical shift in that reasoning.[42]

What Schmetterling proved was that even with all the metals con-tamination and other human changes to the Blackfoot and Clark Fork, the rivers should have supported far more native fish. And the dam was to blame. The dam's spillway allowed fish like cutthroat and bull trout to swim downstream and breed. But they never returned. Schmetterling "ultimately estimated that two hundred thousand fish had their migra-tions annually impeded by the dam." In the fall of 1998, David McEnvoy, one of the UM graduate students that Schmetterling was mentoring, told the local newspaper, "We estimated there were 8,000 suckers in the pool one day. It was kind of disgusting. You could hardly walk across the pool there were so many of them stacked up in there." The urge to spawn was driving those fish to the dam's base. Essentially, the dam was diminishing the reproductive capacity of the Blackfoot and Clark Fork well into their headwaters on the Continental Divide. Whereas the Clark Fork provided warmer water in which many fish overwintered, the Blackfoot offered the best cold-water spawning grounds. Only the resident fish that spawned and bred above the dam kept the upper rivers' population viable, if much re-duced. Of the large number of fish that migrated below the dam, Schmet-terling was amazed that even after ninety years of running up against this impediment, those fish still followed their evolutionary instinct to return upstream past the structure. "One brown trout's been around here for three months trying to get over the dam. Almost every time we come out here we catch him again," marveled Schmetterling. Similarly, many of the fish he observed stuck below the dam had abrasions all over their bod-ies from launching themselves repeatedly against the nearly forty-foot-tall structure. These were Maclean's "high jumping" trout, failing to clear a nearly century-old obstacle. After fruitlessly flailing against the dam "they probably won't spawn again," Schmetterling said. Bruce Farling credited Schmetterling's work with connecting many of the dots between the dam, fish passage, the Blackfoot's health, and the ESA. Once people understood that the dam prohibited bull trout spawning, it became "an illegal take of a listed species" because the dam prevented the fish from fulfilling "its life history strategy, which is to move upstream and spawn," said Farling. As a result, Farling remembered, "what really pushed EPA over the edge to go, 'removal is going to be a serious option' was that ice event and bull trout. The ice event freaked everybody out and then bull trout was sort of an issue that was icing on top of that." By the late '90s everyone who was paying attention knew that the fish in the "place of the big bull trout," as well as hundreds of thousands of other native trout, should be populating Norman Maclean's river.[43]

All those fish piling up under the Milltown Dam recast the understanding of the structure's effect on the rivers' fisheries and strengthened the argument for fish passage. But it was the fish that made it past the dam that really piqued public interest. With the hopes of connecting communities to his research, Schmetterling started an "Adopt a Trout" program. He figured that "the best way to discuss these issues . . . and invite people to see what we're doing . . . was through the eyes of children. . . . It really changed a lot of people's perceptions." By 2001, Schmetterling had about a dozen elementary and high schools in the Blackfoot and surrounding areas tracking native fish that he had implanted with radio transmitters. He gave students weekly updates on where their fish were moving since he'd helped them over the dam. Delighted with the immediate interest, Schmetterling recalled that "we'd go on field trips to the dam so they could watch their fish being captured and implanted and then we'd . . . look at places where those fish are spawning. . . . People got to learn about the history of the dam. . . . Most . . . had never been to the dam."

The Seeley Lake Elementary School, located on the Clearwater Creek tributary in the middle reaches of the Blackfoot Valley, named its adopted cutthroat trout Pollywog. Pollywog found her way up the Blackfoot to Gold Creek, a particularly popular cutthroat spawning tributary. She then returned downstream and hung out in Milltown Reservoir for most of the summer. One day Schmetterling got a call from a man claiming to have caught a radio-tagged pike out of the reservoir. In the late '90s northern pike had moved down into the reservoir from the chain of lakes along the Clearwater River, where they had been illegally introduced. By 1999 fishermen reported catching them regularly in the reservoir, where they thrived in its warm, stagnant water, devouring the native fish. According to Schmetterling, the natives hadn't yet learned to recognize pike as predators, although they were very aggressive ones, especially during spring runoff when cutthroats and bull trout moved downstream through the reservoir at the same time that northern pike went through a "post-spawning . . . feeding frenzy." When he first heard the tale of a tagged pike, Schmetterling didn't believe the man since he hadn't tagged any. The man's father, who was a former director of UM's Wildlife Research Unit, got on the line and confirmed the fish story. When the father and son read him the radio tag information, Schmetterling realized what had happened. The pike and tag were "all that remained of Pollywog," he concluded. The story made the *Missoulian* and, as Schmetterling remembered, "the kids were outraged. The parents were outraged that this cutthroat trout had been eaten by a northern pike. [They] could not believe that people weren't doing something more to stop this." During the course of the "Adopt a Trout" program, Schmetterling encountered many other instances where participants were amazed and angered by how the dam and reservoir were affecting their fish.

Suddenly a whole new contingent of people saw the need for a different solution to Milltown than simply leaving the dam in place. That began to include the MPC. With the number of fish Schmetterling had trapped and counted below the dam, along with the newer concern over pike eating native and threatened fish, "fish passage was going to be really costly," he emphasized.[44]

Schmetterling never explicitly advocated dam removal, but ultimately that was the message people took from his work. Had he simply shown that the dam impeded migrating fish, especially the threatened bull trout, a fish ladder or some other means of transporting native species over the dam might have solved the problem. The combination of the statistics on the large number of native fish the dam halted in their reproductive cycle, the reservoir's provision of habitat for predatory nonnative species, and finally, the expense of transporting fish around the dam made ongoing mitigation seem untenable to an ever-widening community of people.

~ ~ ~

The ice jam of 1996 killed fish downstream of the Milltown Dam. In doing so, it essentially perforated the notion of a contained Superfund site and made the remedy of leaving the dam and sediments in place far less viable. It also helped delay the decision process by forcing a reevaluation of the feasibility study for Milltown. That delay allowed for the unfolding of other events that put the Blackfoot River at the center of the Milltown problem. Schmetterling's work and the specter of having bull trout on the endangered species list "really brought a lot of attention to the dam," according to the biologist. It all happened at a time when the Blackfoot River was in the news and on people's minds. The river's rebounding fishery and the movie that helped revive its popularity and concern for it, along with a raft of study and restoration efforts, all meant that Norman Maclean's river was seeing more traffic than it ever had. By the turn of the century, the Blackfoot had a new set of regulations in place meant to preserve its struggling fish population and manage its human usage. The Blackfoot had also survived a proposal to mine the gold from its headwaters. Federal listing of the bull trout combined with Schmetterling's research made the dam culpable for taking the lives of an endangered species.[45]

In essence, between 1996 and 1998 the Blackfoot River came into the spotlight of the Milltown Superfund process. An ice jam, a major Hollywood movie, the study of native migratory fish, and a fish called Pollywog placed the Blackfoot at the center of the solution for the contaminated reservoir under which it joined the Clark Fork. Those events attracted enough attention to enliven the next major campaign to alter the permanent remedy at Milltown. Following from these events, the first substantial call for removing the Milltown Dam would drown the idea of leaving it all in place. What happened in connection with the Blackfoot River was about to lead to a whole lot more people looking at where Milltown Dam and its reservoir were and considering that perhaps there should be a more natural river running through it.

CHAPTER 6

The Campaign

"Remove the Dam, Restore the River"

Peter Nielsen never imagined that effects from the 9/11 terrorist attacks would reverberate through the Superfund process at Milltown. Yet barely a year after the destruction of the twin towers in New York City, the fear of further terrorism in the United States and the federal government's efforts to prevent it provided one of the last proverbial nails in the coffin of Milltown Dam.

Just before Christmas 2002, Nielsen was sitting in his office at the Missoula City-County Health Department making a routine check of the Federal Energy Regulatory Commission's website. To his surprise, a seismic study of the dam popped up. It described large cracks, massive leaking, and the structure's vulnerability to earthquakes. Nielsen called FERC for an explanation. As he put it, "somebody didn't get the memo at FERC and they were posting documents that were supposed to be classified . . . about the . . . vulnerabilities in our nation's infrastructure." By the next day a TV journalist from a local station visited Nielsen's office, hoping to film the documents on FERC's website. With cameras rolling, Nielsen logged on to the FERC site only to find that the agency had pulled the information, labeling it classified under the auspices of "national security." As Nielsen later learned, part of the federal government's efforts to prevent future 9/11-like attacks enabled FERC to "put the clamps on certain things so terrorists could not get in there and get diagrams to Grand Coulee" dam, for example.[1]

REMOVE THE DAM. Help clean up Milltown Reservoir. **RESTORE THE RIVER.** *www.clarkfork.org*

In the case of Milltown, Montana's tradition of citizen involvement diverged from its usual goal of preservation. Unlike campaigns in support of protecting clean air and water, wilderness, and wildlife, this effort was pushing for restoration. In 2000, the Clark Fork Coalition, a science-based advocacy group focused on water quality, launched an advertising campaign hoping to generate public support for removing the dam and restoring the river, as noted on this bumper sticker. *Courtesy Clark Fork Coalition, Missoula, Montana.*

The story of FERC withholding safety information about a seriously compromised dam that held back contamination in part of the country's largest Superfund site made national news. A number of Freedom of Information Act watchdog websites covered the story. In Missoula, it prompted County Commissioner Barbara Evans to call the dam's owners and convince them to release the reports. "At the end of that conversation," according to Nielsen, the MPC "decided they'd rather take their chances with terrorists" than with Evans. FERC issued Evans a formal apology and agreed to provide the city and county with any information they wanted about the dam. And, the incident helped push Montana's Republican governor Judy Martz to come out in support of dam removal. The mistake at FERC, which associated the Milltown Dam with 9/11, dovetailed with a local campaign that had dramatically raised awareness about the Milltown Superfund site. The result was a flood of public support for the removal of the dam and the site's contaminated sediments.[2]

More than anything, the unlikely contingency of a FERC employee mistakenly posting a classified document about the Milltown Dam and Nielsen happening to see it while "dink[ing] around on the computer" during his lunch hour raised people's fear of leaving an old and damaged dam in place by associating it with the national fears surrounding terrorism. Legitimate fear of keeping the dam in place was only one of the reasons that the Milltown site garnered a record number of public comments during the EPA's efforts to pick a cleanup remedy. In the first few years of the new millennium, people bombarded the EPA with comments about the site because of their fears, their economic concerns, and their sense of fun. Those comments reflected and reiterated the Clark Fork Coalition's advocacy campaign about the site. The CFC's campaign to remove the dam and restore the river transformed the Superfund process. The campaign and resultant public outcry combined Superfund's original mandate to remediate contaminated places and its new dimension of restoring environments.

~ ~ ~

In 1988, Tracy Stone entered the Environmental Studies Program at the University of Montana. She fell in love with Missoula on her first visit and decided, "If I get in, I'm coming. If I don't, I'm coming." In addition to a passion for the place, she convinced herself that environmental studies would be a perfect way to put her undergraduate degree in media studies to work. As she put it, people had "been duped into over-consuming with advertising, perhaps we can be duped through advertising into living more sustainably." Her master's thesis, "Into the Heart of the Beast: A Case for Environmental Advertising," explored this new mode of environmentalism.[3]

Following completion of her degree at UM, Stone worked for various conservation organizations and became a board member for the Clark Fork Coalition, as well as marrying environmental journalist Richard Manning. In 1999, when the director's position opened, she applied. One of her terms for hire was that the organization "expand and change how we do conservation." In particular she wanted the CFC to "be about community" as much as it was "a science-based advocacy group about water quality in a river." The board embraced her vision.

Stone-Manning began her directorship by making the Milltown Dam the coalition's signature project. She asked, "Why aren't we advocating for cleaning it all up and taking out the dam?" By February 2000 the coalition had launched a campaign called "Remove the Dam. Restore the River."[4]

When Stone-Manning met with creative director and copywriter Sean Benton about doing an advertising campaign about Milltown, she knew the CFC couldn't afford much. One of the Missoula-based marketing firm's core principles was that its work should "make the community that we live in better." After hearing about the campaign, Benton and his colleagues at Partners Creative offered to do the work pro bono. Doing free work allowed it the sort of creative freedom that could help distinguish Partners Creative in the competitive ad business. Being proenvironment coincided with growing the business.[5]

The new campaign used multiple media to educate people about the major issues surrounding the Milltown Superfund site and the forthcoming decisions about its cleanup. It drew on support from around the community, especially small businesses, and it was meant to be fun, even funny. Stone-Manning hoped that balancing humor with serious issues would entice people who didn't consider themselves environmentalists to support an environmental issue. It worked. The campaign got people involved in shaping the outcome of federal public policy by educating them with advertising. Most of that education centered on recreational interests, the idea of restoration, and the fears surrounding Milltown's Superfund site.

People who had been paying attention to the Milltown site already had fears about it before the CFC's campaign. So it wasn't surprising that the advertising effort drew on those fears. The campaign and people's responses centered on fears about the dam's safety, the toxins it withheld, and the responsible companies.

Fear of a catastrophic failure wasn't new in 2002, when FERC accidentally posted information about Milltown Dam's instability. Locals had feared the dam would collapse as far back as the flood of 1908 and as recently as the ice-scouring event of 1996. The CFC dam removal campaign reiterated such fears. In 2001, one of the group's local TV advertisements began with a soundtrack reminiscent of Darth Vader's breathing, followed by a series of dissonant, industrial tones. The ad's only visual was a white liquid infiltrated by a surge of black ink that swirled and filled the screen as the narrator spoke about the "poisoning" and "threatening" effect of the Milltown sediments. Newspaper advertisements included the image of a skull and crossbones with the word "POISON" underneath.

The Clark Fork Coalition used the phrase "Not All Time Bombs Tick" in numerous newspaper advertisements. Its most prominent use was on a billboard along the highway between Missoula and Milltown. The billboard featured a black-and-white photo of the Milltown Dam's downstream face—the view from Missoula. The dam's spillway and gates were caked in ice and snow, as were the wooded mountainsides in the background, which was a visual reminder of the 1996 ice-scouring event that nearly destroyed the structure. Newspaper ads equating the dam with

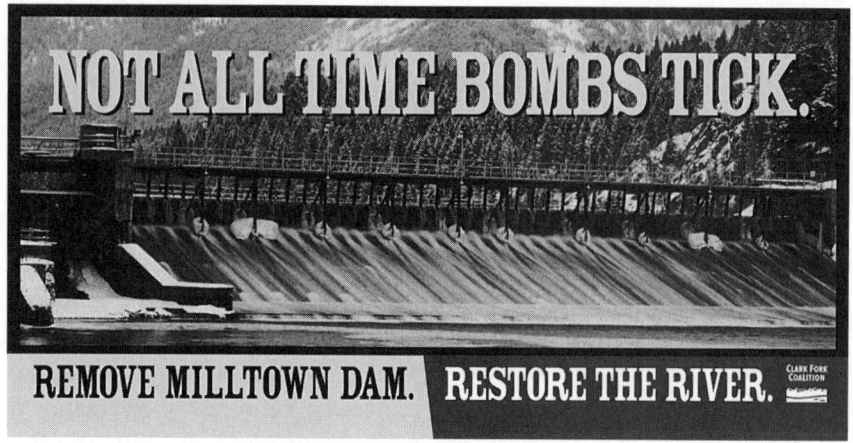

NOT ALL TIME BOMBS TICK.

REMOVE MILLTOWN DAM.   RESTORE THE RIVER.

One of the Clark Fork Coalition's signature messages, which headlined news-paper advertisements and appeared on a Missoula billboard, read "Not All Time Bombs Tick." A 2001 *New York Times* article reported that local residents "all along the political spectrum" were beginning to "portray the dam as a toxic time bomb." One study found that sixty-eight thousand tons of metal-laced sediment flowed through and over the dam every year—a steady trickle compared to what a flood-wrecked dam might unleash. *Courtesy Partners Creative and Clark Fork Coalition, Missoula, Montana.*

a bomb threat were a potent post-9/11 symbol. Referring to the Milltown Dam as part of "one of the largest Superfund sites in the nation," a 2001 *New York Times* article, like most CFC ads, detailed the contents of the proverbial bomb in terms of heavy metal quantities and the vulnerability of the dam.[6]

When it came to playing on fear of toxins, Milltown's problems made it fairly easy to create copy. "Arsenic is bad. Had it been something that nobody knew [about] behind the dam, it would have been a lot harder. But everybody knows arsenic," remembered Benton. The campaign tried to turn the figure of 6.6 million cubic feet of contaminated sediments into local knowledge as common as the ounces of soda in a can. In the buildup to the EPA's final decision on Milltown, virtually every letter the agency received in favor of dam removal addressed the tandem worry of an old, unsafe dam and the toxins behind it. Many quoted the statistics the CFC campaign made so widely known.[7]

Whereas environmental realities made the campaign's message about toxins accessible, timing did so for its focus on fears about the responsible corporations. During the first few years of the new millennium, as the EPA and ARCO worked through a second and third feasibility study necessitated by public and organizational input, the potentially liable parties began changing hands. As early as the summer of 1997 the Montana Power Company was trying to rid itself of the dam. Within a year, the MPC was selling its hydropower plants and trying, as the company vice president put it, "to get out of the power generation business." The company knew that it would be hard to sell the dam because of its low power output and entanglement in Superfund. The Missoula city-county government and other observers, like the CFC, worried that if the MPC sold the dam the trail of liability would become longer, more complex, and more tenuous. In November 1998, the MPC sold all its power-generating facilities except the Milltown Dam to PP&L Global, just as ARCO was finishing a Focused Feasibility Study, a revision of the original FS that the EPA had demanded because of people's unsolicited comments indicating that they "didn't believe that the dam removal option was given proper consideration."[8]

The MPC's financial role at the site was uncertain, whereas ARCO's appeared integral to the possibility of a polluter-pays cleanup. When BP Amoco sought to buy ARCO for $25 billion in March 1999, EPA spokesperson Pam Hillary had to assure Montanans that the new owners "won't be off the hook."[9] In fact, the purchase, which created the world's second-largest oil company, would eventually facilitate the removal of the Milltown Dam. At the time, the worry was that the liable company at the site would become even less amenable to environmental cleanup. As the sale was happening, ARCO settled with the state of Montana's Natural Resource Damage Program and the Confederated Salish and Kootenai Tribes over natural resource damage claims for $260 million. Culminating a sixteen-year court case, the settlement was only about a third of the $765 million the state had once sought. Advocates for river cleanup thought that the money was a boon for the Clark Fork's future, but also a sign of ARCO's intention to pay for as little of that future as possible.[10]

In July 1999, BP and ARCO sealed their deal. That deal included, as the *Missoulian* front page reported, a $63 million severance package for the seller's "top five officials."[11] In the same month, the EPA received a draft of ARCO's Focused Feasibility Study (FFS).[12] Upon reviewing it, the agency forced the company to make two major changes. Both pointed to the company's preference for minimizing its costs. First, according to the EPA, ARCO had underestimated cost information for dam safety improvements and fish passage construction for the option of leaving the dam in place. Second, ARCO had overstated the risks of sediment and dam removal, thereby making that option appear more risky and expensive than the agency deemed it. ARCO preferred a solution that would cost less than one out-going executive's severance package.[13]

ARCO increased suspicion about its commitment to cleaning up the dam site the following spring. In the midst of another round of license extension requests from the Montana Power Company to FERC, which nearly everyone who was paying attention opposed, ARCO struck a private deal with the dam owners.[14] The MPC reported that it had agreed to join ARCO in support of a "dam-in-place and sediments-in-place remedy for the Milltown Reservoir." Calling it a "Stewardship Agreement," the companies committed to "search for a third party to assume local control of the dam into the future" and to create an operating fund for the prospective owner. The deal promoted shoring up the aging dam with an "inflatable . . . rubber structure" along its crest, leaving all the toxic sediment behind it, and then handing the whole thing over to the locals.[15]

As the *Missoulian* reported in early May 2000, "the response was quick and critical." It was also fearful. Local government agencies worried that the deal would "leave a mess for us to deal with in this community." Peter Nielsen saw the move in the context of Superfund sites nationwide, which he worried were getting incomplete remediation and being left for local governments to wrestle with. Since the Clinton-era Congress had failed to maintain the taxes that targeted industry and funded such "orphan sites," those sites were receiving ever-decreasing support from Superfund. While private industry had paid for cleanup at roughly two-thirds of Superfund sites since its enactment, funding for the other third "dwindled from a high of $3.8 billion in 1996 to a projected $28 million" by 2003, according to the *New York Times*. The Bush White House made no effort to reinstate a corporate tax to support Superfund sites without liable parties. As Superfund shrank, orphan sites depended increasingly on general appropriations from Congress for cleanup money. At Milltown, Nielsen saw an orphan site in the making. "The further away you get from the companies . . . the less likely it is that you'll get anything out of them . . . ARCO is being swallowed up by BP-Amoco and Montana Power is going away. And now they're talking about turning the whole thing over to a nonprofit group. Get real." Seconding Nielsen's unease, the Missoula county commissioners released a report called "Case for Removal of Milltown Dam," which expressed fears that the community would take on the burden of future problems if the dam stayed and its former owner disappeared.[16]

Along with keeping attention on dam safety concerns and toxins, the Clark Fork Coalition campaign kept the responsible parties in the spotlight. All of the coalition's newspaper advertisements that encouraged public participation summarized corporate responsibility. One ad reminded readers, "EPA has requested a cleanup proposal from ARCO, the company liable for this mess. But this subsidiary of multibillion dollar conglomerate BP seemingly intends to do as little as possible." Another ad pointed out the small financial burden that dam and sediment removal would entail, and hence the pettiness of corporate refusal to embrace that option. It estimated that "the entire cleanup would . . . cost $100 to $150 million. That's a lot of money to us, but it's only about 1 percent of the net profit BP made in 2000." The spring 2001 edition of the CFC newsletter, *Currents*, reiterated that point with the metaphor, "For this company, $130 million is the household equivalent of, say dinner and a movie."[17]

The public's reaction to the corporations reflected the CFC dam removal campaign. All its campaign ads encouraged people to contact the EPA and express their support for dam and sediment removal, as well as restoration.[18] People responded. They often did so by reflecting the campaign's language of fear. In the end, the federal agency received about ten thousand public comments on Milltown, which was a Superfund record.[19]

Many commenters charged the company with both past grievances and obstruction of the current cleanup. Voicing his hopes for dam removal in an e-mail to the EPA, Missoula resident John Adams opined, "When I hear company reps whine about the cost, I can only think back on the billions of dollars of profits they took out of Montana in creating this problem." Catherine Price-Alick and Claude Alick e-mailed an even more vitriolic message to the EPA, equating Milltown's history with "big greedy corporations" that "have stolen Montana's wealth for years, killing people, destroying the earth, and then leaving us, poor taxpayers who didn't 'benefit' from the theft, to clean up the toxic waste ourselves."[20]

Sometimes the influence of the local environmental campaign was even more obvious. Numerous people quoted campaign literature verbatim when they equated BP's ability to pay with a date-night expenditure. At the same time, the companies tried to allay people's fears with a few of their own.[21]

ARCO's president of environmental remediation, Sandra Stash, voiced the company's worries about dam removal. She claimed to be "frustrated by that whole discussion," calling the idea of taking down the dam "impractical." In 2001 Stash told the *New York Times* that removing the dam would "do more harm than good." Penning the company's official response to the Missoula County resolution supporting dam removal, Stash summarized that prospect as carrying "extremely severe practical problems and risks." Similarly, Montana Power spokesperson Jim Williams equated dam removal with "crossing an interstate blindfolded." Their attempts at raising counterfears seemed to pale compared to the "ticking time bomb." When details of the closed-door deal between the MPC and ARCO came to light, along with reports that the MPC had hidden more information about leaks in its dam, fears of a superficial cleanup seemed justified.[22]

Because of public pressure and the Clark Fork Coalition's use of Montana's sunshine laws, ARCO and the MPC had to release the details of their agreement for public scrutiny in early 2002. Their deal limited the dam owner's liability to $10 million, as long as the MPC coordinated all of its efforts in support of ARCO's cleanup choice, which was the least expensive method of keeping the dam and sediments in place. ARCO's threat to sue the MPC explained why the power-turned-telecommunications company had abruptly switched from trying to decommission the dam to applying for another series of license extensions. Apparently the MPC saw the deal with ARCO as a safer and cheaper bet than relying on the 1986 Superfund Amendments and Reauthorization Act provision that tentatively released the dam owners from liability. As CFC attorney Matt Gifford gleaned from the legal provisions of the deal, "when Montana Power says it wants to keep the dam" even though it's a money-losing prospect, "that's ARCO talking."

What ARCO was talking about was exactly what the coalition and local government had feared—spending as little money as possible and walking away. In each of its three feasibility studies the company did its best to favor dam retention, whether that meant distorting the expense and its ability to implement cleanup, or simply offering its opinion and coercing the MPC into following suit. In the long and tedious process of writing a feasibility study acceptable to the EPA, the company faced the steadfast scrutiny of local environmental advocates, an equally engaged local citizenry, and the press.[23]

As with many turns in the Milltown story that revealed the details of Superfund implementation, public input was key. And in the case of the fears surrounding corporate buyouts and liability, environmental marketing helped inform public input, which flooded EPA project manager Russ Forba. Along with individuals, organizations petitioned the agency. The CFC rebutted many of the findings in ARCO's FFS, especially its underestimation of risks and costs associated with leaving the dam and toxins in place, as well as its inflation of costs and risks linked to dam and sediment removal. The Missoula City-County Health Department submitted the exact same complaints to the EPA. The Clark Fork River Technical Assistance Committee, the Superfund-funded information outlet for the site, trebled those observations, explicitly stating that "the analyses are cursory and subjective, aimed at finding the cheapest alternative without regard for what is best for the river." Montana Fish, Wildlife and Parks agreed. Arvid Hiller, the vice president and general manager of Mountain Water Company, which provided Missoula's drinking water, feared that the remedies outlined in the FFS provided insufficient safeguards for the downstream sole-source aquifer. The U.S. Department of the Interior's Fish and Wildlife Service added its fear that an in-place solution would fail to protect the threatened bull trout. The U.S. Army Corps of Engineers commented that ARCO's draft FFS underestimated the "implementability" of removing sediments.[24]

At the behest of the EPA, the Corps of Engineers conducted studies of dredging scenarios for the Milltown sediments and found that they amounted to common earth-moving ventures, which carried little danger of polluting the river or downstream aquifer. Based on its growing experience with dismantling such obstructions, the Corps returned a much lower cost estimate for dam removal than ARCO. By April 2001, the EPA engaged ARCO in a thorough rewrite of the draft FFS, especially with regard to cost estimates and the "implementability" of various options.[25] The final product, a "Combined Feasibility Study," the last assessment of cleanup options leading to an EPA decision at Milltown, analyzed costs and downstream impacts more rigorously. And it, too, lessened the risk of implementing dam and sediment removal.[26]

Although the EPA's official public comment period was supposed to follow the release of the Combined Feasibility Study, people continued to speak up ahead of schedule, and they repeatedly drew on the environmental campaign's language to do so.[27] Fear of corporations surfaced most consistently. Rosanne Davis lamented that if the EPA failed to enforce a full cleanup, including dam removal, "the people/taxpayers end up being responsible for the arrogance and greed of the mining industry." She ended her letter by scolding the EPA that "to continue to dance to the tune of the mining industry and gambling with our health is unconscionable. It's about time that Atlantic Richfield took responsibility for the wealth it extracted from Montana and the destruction it left in its wake." Although Davis, like many other people, was confused about ARCO harboring liability not because it had done the mining and polluting but because it had bought the company that did, her anger at the past and fear of the future regarding corporations was commonplace.[28]

With MPC contractors using sand and grout to plug cracks in the dam that measured up to four inches wide, with the dam up for sale and the MPC getting out of the power business, with ARCO and the MPC coordinating their efforts to back a minimal cost and minimal cleanup option, with ARCO being folded into the world's second-largest energy company, and with the site's risk assessments reporting plenty of problems with leaving the toxic sediments in place, perhaps the focus of public input on fear was not surprising. What *was* surprising was a seamless blend of fear with fun in the dam removal campaign.[29]

~ ~ ~

Many people know the environment through play. That was certainly true for Sean Benton. He first forayed west from Kansas City on overnight drives to Colorado, leaving his marketing job on a Friday so that he could lash his fly rod at mountain streams first thing Saturday morning. Moving to Missoula and getting Partners Creative started in 2000 helped shorten his drives from business to pleasure.

Thinking back about the advertisements he helped design for the Clark Fork Coalition, Benton mused, "It's a commercial world and that's the way it works. . . . I'm not a big fan of seeing a billboard on my way to Glacier National Park either, but they're there. They can have a message about Miller Light or they can have a message about Remove the Dam. The billboards that we did wound up being some of the things that are most recognized."

Along with the toxic time bomb billboard, another that attracted lots of attention read "Ski Milltown! It's Toxic!," injecting a serious issue with a dose of humor. *Courtesy Partners Creative and Clark Fork Coalition, Missoula, Montana.*

Like Stone-Manning, Benton believed that commercialism had a role to play in environmentalism beyond making products seem green.[30]

Along with the toxic time bomb billboard, the one that got the most attention read "Ski Milltown! It's Toxic!" If the message was surprising, the image it overlaid was more so. In shades of green the color of an algae-choked lake, the billboard featured four smiling women clad in matching tutu-style bathing suits and tiaras. The quartet appeared to be water skiing barefoot with their right legs tucked under them like flamingos and their right hands placed on the next skier's shoulder in a chorus-line pose. The image conjured a 1940s Palm Beach water show far more than it seemed akin to anything that might unfold at Milltown. It was also a new tactic by the Clark Fork Coalition.[31]

The message dutifully referenced the real toxicity of Milltown. Underpinning the "It's Toxic" text, the color of the billboard spoke of polluted water in a watershed prized for its clear, cold trout streams. Water skiing meant recreation, although the image made light of that particular form of play in a river environment. The reservoir itself was out of place by implication. Most importantly to the ad's creators, it was funny.

The coalition retained its emphasis on the serious problems at Milltown by keeping toxicity in its ads, which was especially effective since the EPA passed a more restrictive law governing arsenic in drinking water in the midst of the dam removal campaign.[32] In May 2000, the *New York Times* announced the Clinton administration's proposal for the EPA to lower the amount of arsenic allowed in drinking water by *ninety* percent. Within a month, the Clark Fork River Technical Assistance Committee (CFRTAC) wrote to the EPA about how a change in arsenic standards might "affect the plume boundary in the Milltown area." If far less arsenic in water posed a health risk, then far more water supplies would be considered unhealthy.[33] Upon taking office, President Bush promptly suspended the Clinton initiative, which began months of public reaction and agency pressure. He recanted eventually, just as he came to support Milltown Dam's removal after having delivered a campaign speech in 2000 supporting retention of Snake River dams that faced environmental challenges because of their obstruction of salmon passage, as well as using the opportunity to promise that "removing dams would not be on the table in a Bush presidency," as the *Seattle Times* paraphrased it.[34] In February 2001, the *Missoulian* ran a front-page article on the EPA's decision to abide by new research showing that ingesting even small amounts of arsenic increased people's risk of cancer and other deadly ailments. CFC director Stone-Manning connected the national shift in environmental regulation with the local issue, observing that "EPA's tougher stance on arsenic pollution is just one more reason why Milltown Reservoir needs a radical cleanup."[35]

Nearly every campaign ad quantified the reservoir's toxic load. One used the headline "Tons of Incentive to Speak Out" to point out that "19 tons of arsenic" flowed downstream annually. A coalition-sponsored local TV clip showed underwater footage of a pair of trout swimming upstream while a male narrator described the poisonous water that was killing those fish as long as the dam stayed in place. The message about heavy metals was clear.[36]

The use of water skiers and trout in these advertisements was meant to speak to recreationists as much as anyone. The research Benton and Partners Creative did before designing their advertisements proved that the watershed was home to plenty of people who cared about the river's fishery because they fished for sport. Between efforts to clean up tributaries on the Blackfoot; Redford's film version of *A River Runs through It*; FWP fisheries biologist David Schmetterling's work on fish passage at the Milltown Dam, including the story of a reservoir pike eating the trout Pollywog; and the Department of the Interior's listing of bull trout as threatened, both local and out-of-state people who fished were paying attention to the Superfund process in one of western Montana's most coveted recreational watersheds.

Anglers weren't the only ones thinking about recreational opportunities in association with remediation at Milltown. By early 2001 the Missoula county commissioners had drawn up a proposal called the "Two Rivers Restoration and Development Project." The proposal looked past dam removal to "environmental restoration and recreational development." The county's "once in a lifetime opportunity to do something very special for our community" included an extended nature trail, footbridges, a "Riverside Park," and "construction of whitewater features on both the Clark Fork and Blackfoot Rivers."[37] The *Missoulian* ran a feature on the proposal in which one of its designers, Gary Lacy, envisioned how "this project would bring the rivers back into . . . a natural appearing whitewater stretch that is attractive for all the reasons why people like rivers," such as "deep pools for fishing and paddling . . . waves and drops and white water." Those same features would benefit fish, too, but recreation and outdoor fun were helping to sell the idea of dam removal and restoration.

Sporting organizations like the Montana chapter of the American Fisheries Society, the American Whitewater Association, and the Missoula Whitewater Association, as well as owners of local sporting goods businesses and many Missoula residents, approved of dam removal. Missoula Whitewater Association president John Anderson wrote to contribute his members' support and to send the EPA copies of a *Paddler Magazine* article about the popularity and economic success of whitewater parks that western towns and cities had created in recent years. Brian Daly, the owner of a local software development company, wrote to EPA state director John Wardell that "although I am one of the staunches [*sic*] Republicans you'll ever find . . . the plan for a 'Whitewater Park' at the confluence of the

Clark Fork and Blackfoot rivers makes a lot of sense to me." Of the thousands of letters supporting dam removal that the EPA eventually received, many mentioned recreation.[38]

In fact, the demographic of river recreationists was prevalent enough that Partners Creative and the coalition opted *not* to make harm to recreational fishing an explicit part of their campaign. Simply having fish in their ads and talking about the detrimental effects of Milltown to the fishery would attract the constituency that Benton identified as "forty plus and they're male and they like to fish." In other words, the ads only had to hint at the recreational aspects of the rivers to attract the support of fisherfolk, boaters, or riverside hikers. The "Ski Milltown!" billboard implied, "in a very cheeky way," according to Benton, that the reservoir was "not a flat-water resource" worth keeping. Rather, it was something that could be a true river recreation resource.[39]

Stone-Manning ventured that the "cheeky" aspect of such ads would draw unlikely support for the campaign. Tired of seeing environmental groups fighting *against* things or always saying "no," she wanted to advertise the positive. She also chafed against environmental organizations' tendency to "preach to ourselves." With the goal to "get people excited" about the dam removal, humor became one of the campaign staples.

When Stone-Manning saw the first samples of the "Ski Milltown!" image, she thought "it was funny!" Wanting to gauge its appeal to a new set of people, especially nonenvironmentalists, she began showing it around. When a very conservative county auditor laughed at it, Stone-Manning asked, "You don't think it goes too far?" The woman responded, "That's the problem with environmentalists. They don't laugh enough." With that, Stone-Manning knew it was time to run the billboard. As she recalled, "I want to capture this woman and all of her friends."[40]

A sense of fun about this serious environmental issue peaked when the Clark Fork Coalition sponsored some actual playtime on the river. The temperature reached 104 degrees in Missoula on July 13, 2002. Sweltering summer heat on a Saturday meant heavy traffic on the local rivers—families and college students floating in tubes, anglers rowing rafts, and the occasional lone kayaker. But temperature and weekend free time could not have drawn the number of people drifting toward Missoula on the Clark Fork that day. It was the work of the coalition's campaign to remove Milltown Dam.

A sense of fun peaked when the Clark Fork Coalition sponsored some actual playtime on the river. In July 2002, the coalition advertised the first "Milltown to Downtown" float, and more than 150 craft and an estimated 500 or more people showed up on the water, demonstrating what was by then overwhelming public support for removing the dam. By Halloween, the EPA had counted more than 10,000 public comments on Milltown—an "unheard of" response outside a formal public comment period. *Courtesy Clark Fork Coalition, Missoula, Montana.*

"Ideas came from the community," Stone-Manning remembered about the origins of some aspects of the campaign. Her favorite was from a CFC member. Jack Moyer spotted Stone-Manning's tall, thin frame and curly red hair in a grocery store, where he walked up to her and said, "Do you know what you need? You need a big flotilla from the dam into Missoula!" "We totally need a big flotilla from the dam into Missoula," she responded. So the coalition advertised the first "Milltown to Downtown" float in support of removing the dam. They *attempted* to organize it, anyway.

CFC staff rounded up about fifteen friends who had boats to accommodate the dozens of people Stone-Manning expected to show up. By the time temperatures were peaking, more than 150 craft and an estimated 500 or more people were on the water. And that didn't include the scores of dogs. "It was crazy! It still makes my eyes water," confessed Stone-Manning. She recognized only a fraction of the people there. She thought, "Who are these people[?] . . . I love them." It was, according to her, "not those environmentalists from Missoula . . . it was grandmas, babies, and frat boys . . . it was a very cool expression of people saying, 'this is what we want!'" Some wanted dam removal because of fish or drinking water concerns, others so that they could recreate more safely, and some because of the economics of recreation. Whitewater guide Morgan Valliant warned, "Do not underestimate the economic force of recreation . . . taking the dam out of this river would be huge because it would restore the native fish migrations. Fishing guides would really benefit." As Stone-Manning's raft cruised into Missoula she looked up at one of the bridges and saw John Wardell, the EPA's state director, "with a big grin on his face." She knew that he saw the same thing she did, overwhelming public support for his agency's option of removing the dam. County Commissioner Barbara Evans got out of the river slightly wet, but her excitement was far from dampened. "This ought to show the community concern," Evans quipped to a reporter. As the *Missoulian* reported, the only opposition present was a riverside sign imploring "Keep the Dam, Kill the Hippie." To Stone-Manning and Benton, seeing the opposition riff on their slogans, especially in a funny way, confirmed that the campaign was really working.[41]

The second annual Milltown to Downtown event drew nearly three thousand people. More importantly, the EPA received an estimated ten thousand public comments about the Milltown decision. It was this flood of public comment that helped sway the EPA. People's comments reflected fear of leaving the dam in place as well as fear of being left with the cleanup bill by the responsible companies. They spoke of recreational connections to the Clark Fork and Blackfoot and the recreational potential of the watershed. Many reiterated human and environmental health concerns. In addition, they nearly all linked dam removal with restoration.

~ ~ ~

Diana Hammer saw Milltown for the first time in 2001. Compared to other Superfund projects on which she'd worked as the EPA's community coordinator, Milltown struck her as "really beautiful." Yet Hammer knew that when it came to the beauty of some Superfund sites, especially the ones she'd seen in Montana, "your eyes can fool you." She also knew that it was the CFC campaign, along with the EPA grant–funded Clark Fork River Technical Assistance Committee, that "did a ton of leg work" to get people involved in the Superfund education and comment process. Those comments, according to Hammer, prompted the EPA to make dam removal one of two final cleanup options. As a lifelong student of the environment and a river rafter, Hammer felt it "was a really quite exciting time." In addition to personal interest, she knew that Superfund had no history with dam removals. Milltown, if it happened, would be a first.[42]

The Milltown Superfund site also gained a historic reputation for public input. At Milltown, as with any Superfund site, the EPA had guidelines for public comment periods. Probably the most important of those periods surrounded the feasibility study, or in the case of Milltown, the Combined Feasibility Study (CFS), which public input had already altered outside the confines of regular comment periods. After the release of any FS, the EPA solicits "public review." This review and the input of the site's state government (essentially the governor's office) are the last two significant factors the EPA uses to choose a cleanup option.[43] A 2001 study of Superfund implementation showed that in making cleanup decisions, the EPA prioritized the involvement of "concentrated private interests, such as liable parties and local communities," over even costs or contaminants.[44] A similar 2003 study suggested "the importance of state-level factors in Superfund implementation."[45] At Milltown, people and organizations had submitted comments throughout the CFC dam removal campaign, well before the release of the CFS. And they did so in record numbers. Those numbers helped sway the state.

The campaign encouraged pushing Superfund beyond remediation and toward *restoration*. As such, Milltown was at the crest of the wave of environmental restoration that signaled a new direction in American environmentalism. People have restored ecosystems for hundreds, if not thousands, of years. For millennia people have used fire to regenerate landscapes. In the United States there have been successful efforts to restore endangered wildlife populations dating to the late nineteenth century. Yet it wasn't until the 1970s that "land managers, conservationists, and environmentalists" began to consider systematic ecosystem restoration as a land management practice. Individual practitioners of ecosystem restoration helped birth the journal *Restoration & Management Notes* (later *Ecological Restoration*) in 1981 just as the Superfund law resulted in the first National Priorities List of federally designated toxic waste sites, including

Milltown. According to historians of the earliest incorporation of restoration into American environmentalism, "this constitute[d] something of a revolution in environmental thinking and practice." And, a "watershed in this development was the passage in 1980" of CERCLA.[46]

Watersheds were at the forefront of the new emphasis on restoration in American environmentalism. As historians of that new emphasis have argued, "stream restoration, in particular, has become a flagship for the restoration movement, linked to a range of issues from water quality to endangered species to recreation and drawing the lion's share of public attention." The EPA reported that by 2010 there "were more than 2,600 adopt-a-watershed groups nationwide," a number which, according to studies, was "orders of magnitude higher than for other types of restoration." Local groups concerned with the health of their water led the way. Long before 2010, the CFC was doing just that.[47]

The original Superfund Act said nothing about restoration, but on the Clark Fork its implementation was making it an intended outcome. Restoration was not a new idea at Milltown. In 1999 the state's Natural Resource Damage Program won a $260 million lawsuit against ARCO, which would help fund "recovery" of the Clark Fork environment. But the CFC campaign linked river restoration to dam removal. Every billboard, print ad, and TV spot included the paired commands: "Remove the Dam. Restore the River." "You couldn't just take down the dam and see what happens," said Sean Benton of the decision to link toxic cleanup and restoration. As the mantra of restoring Milltown gained momentum, people embraced a very practical definition of restoration, which included a free-flowing river system and reconnected trout migration routes, as well as "redevelopment" of the local economy. Specifics of what restoration actually meant developed only after the decision to remove the dam was made.[48]

Removing the dam as a catalyst to developing new financial opportunities in the Milltown area was, in large part, a product of opposition to dam removal. Just as remove-and-restore advocates were winning overwhelming public support, opponents of that option shaped the debate about what would happen to local people if the rivers ran free. As such, Milltown became an example of a new direction for Superfund and environmentalism, as well as a case study in how struggles over the environment are not a winner-take-all prospect. As such, the "really beautiful" Superfund site that Diana Hammer saw was the site of that law's first dam removal as well as its move toward restoring environments and stimulating economic redevelopment for the communities that surrounded them.

~ ~ ~

As EPA's project manager for the site, Russ Forba, confirmed, "It's not a vote, but what the public says will influence what we say." Hoping that the public's say would be heard, CFC board president Geoff Sutton encouraged, "If the community yells loud enough, FERC and EPA will listen. But we've got to yell."[49]

The "yell" of support for dam removal that Sutton sought came in many forms. E-mails and letters addressing Milltown continued to arrive in the EPA's state office. So did the results of a few community studies. In early 2001, the University of Montana's Sociology Department conducted a survey of Missoula County residents that revealed that 68 percent of people surveyed supported full cleanup of the reservoir, with a strong correlation between people's level of knowledge about the issue and their support for cleanup. Dam removal supporters began to sway local government. At the end of March, Missoula county commissioners passed a resolution in favor of dam removal in reaction to public input. In editorials, the *Missoulian* threw its support in the same direction.[50]

The county commissioners' support for dam removal brought important bipartisan backing to the issue. Leaders of the conservation community and local agencies involved at Milltown all pointed to the importance of Commissioner Barbara Evans's decision to advocate for dam removal.[51] Evans, a Republican, came to that decision because of her worries about long-term human and environmental health issues, such as a tainted Missoula aquifer. As a fiscal conservative, she was wary of county citizens and county government having to pay for future cleanup needs if the liable corporations funded an incomplete or faulty remediation. When ARCO released the CFS and backed its own plan of placing a synthetic extension on top of the dam, Evans criticized it as the "rubber hot dog" fix. Because Evans was a conservative Republican with a long history in local politics, her support "made it a bipartisan thing right away . . . it became . . . a community issue," said Tracy Stone-Manning.[52]

Montana's Trout Unlimited director, Bruce Farling, agreed about the influence of Evans's support, which he remembered came out of her personal interests in the environment as much as her politics. Farling said of the commissioner, "she's not an environmentalist. But she has . . . affection for frogs. She's really into frogs. When she found out that this contaminated eggs . . . and high metals kill frogs, she started paying atten-tion."[53] In 2000 Evans revealed her environmental interests when she told the *Missoulian*, "I want that reservoir cleaned up, so it is a safe place for people and for fish and for the frogs." Evans and her fellow commissioners wrote an opinion to the newspaper featuring an illustration of three men trying to maneuver a boat through a large crack in a dam. The opinion defended a resolution the commissioners adopted on March 28, 2001, calling for dam removal. The resolution and subsequent opinion invoked bipartisan community support for dam removal and restoration.[54]

The evolving solution at Milltown represented a national trend toward solving environmental problems concerning rivers through a nonpartisan, whole-watershed approach. The diversity of watersheds and people's use of them seemed to necessitate a move away from the divisively partisan atti-tudes about the environment that had flourished in the 1980s and '90s, ac-cording to historians. As Tim Palmer wrote about groups like the CFC in *Endangered Rivers and the Conservation Movement*, "new groups sought to include representatives from all different points of view, each mem-ber a 'stakeholder.'" Once Milltown was designated a Superfund site, no one opposed environmental cleanup. Rather, liable parties sought to limit their financial obligations, and the majority of public participants, and local and state agencies, pushed for thorough remediation. Not whether, but how, cleanup should happen was at issue. Antagonistic political par-ties played little or no part in that decision. Rather, the notion of cleanup teetered between containment—the original Superfund model—and restoration.[55]

With the release of the CFS and the opening of a new public comment period, people were supposed to voice their preference for one of two re-medial options—2A or 7A. They certainly did that. They also pushed be-yond the scope of remediation in favoring restoration and redevelopment of the confluence. A very representative e-mail came from Steven E. Bris-tol, who wrote that as a Clark Fork watershed resident he thought that all the evidence indicated "the dam needs to be taken away and the river re-turned to its natural state." Jeff Rolston-Clemmer agreed that "as a resident of Missoula . . . I think removal and restoration of the Milltown Reservoir is the only safe and permanent solution." Many others concurred.[56]

Meanwhile, plenty of people included economic development, or what came to be called redevelopment, in their opinions. Brad Robinson, part owner of Missoula's Big Sky Brewery, e-mailed the EPA and governor that "the economic benefits to the greater Missoula area will be tremendous." He foresaw that "recreational tourists will travel from great distances to use this new free flowing and clean river" and "will increase the number of people who . . . stay in Hotels and eat in Restaurants," as well as, perhaps, drink the local beer. Labor groups also called for using the removal of Milltown Dam to create jobs. In a letter entreating Governor Martz to favor dam removal, the Montana State AFL-CIO considered option 7A "a potentially huge economic development opportunity for Montana." Whether urging removal because of anticorporate feelings, care for the future, recreation, economic benefits, or hopes for restoration, an increasing number of people wrote to Governor Martz.[57]

The Clark Fork Coalition campaign helped inform people of the necessity of gaining the state's voice in their collective "yell" for dam removal. That meant getting Martz on board, a Republican who had once referred to herself as the "lapdog of industry" and who had a poor environmental record according to the state's conservation community. When the inaugural Milltown to Downtown float took place in July 2002, "community acceptance" of a solution for the Superfund site was pretty clear. By that time the state's EPA office had received "8,312 comments on the Milltown cleanup, a number unheard of for a Superfund site—particularly one that does not yet have a proposed cleanup plan," reported the *Missoulian*. The EPA's Diana Hammer called the "7,801 comments—or 94 percent" supporting dam and sediment removal "pretty impressive, and very unusual." Copies of most of those comments sat on the governor's desk.[58]

On November 5, 2001, the EPA released its Final Combined Feasibility Study. The agency also announced it would propose a cleanup plan by "early 2003." Diana Hammer, the EPA community coordinator for Milltown, told the *Missoulian*, "This is the time for people to tell us what they want. . . . This is when public comment really counts." The EPA was set to hear more public support. It had also become clear who opposed it.[59]

ARCO touted the cost efficiency and innovativeness of the "rubber hot dog" fix, as County Commissioner Barbara Evans called it, while dam removal advocates feared both those reasons—one as a cheap fix, the other as a dangerous experiment where a permanent solution was the goal. The CFS estimated ARCO would spend $20 million on the rubber solution and up to $319 million on dam removal. Outside the company, people who championed keeping the dam tended not to address the rubber dam at all. Rather, proponents of alternative 2A fretted over what they saw as the negative impacts of removing the dam. They, too, spoke in terms of permanence, fear, and environmental and economic concerns.[60]

The Bonner Development Group (BDG) formed in the mid-1990s as a community advocacy organization in the Bonner and Milltown area. Local residents and business leaders came together to help promote and carry out small community-development projects and advocate for their towns within the county government. By 1996 the group considered accepting "financial assistance from ARCO to staff and operate the organization." Jerry Tavegia, one of the state's economic development officers for the Montana Department of Commerce, advised the group that "accepting start-up funding from ARCO" would not lead to compromising the group's independence. Rather, corporate funding would be "a tool to enable the organization to mature." The group's most active leaders, Gary Matson and Bruce Hall, accepted that advice and corporate dollars.[61]

By early 1999, Bruce Hall was the executive director of the BDG, whereas Gary Matson had ceased to support the group. Hall's organization became the most prominent public opponent of removing the Milltown Dam. As Hall articulated it, the BDG opposed removing the dam because for Milltown it was "an important piece of the town's economy." In newspaper articles and letters to the EPA, the group repeated the statistic that the dam contributed "$200,000 — or about 15 percent — of the local school funding base." Taking out the structure would also destroy the "wetlands and recreational opportunities" that it provided. Furthermore, the group's members and their community supporters expressed fear about the damage dam removal could cause to the downstream watershed, or what Bruce Hall called "a Pandora's Box of unexplored and highly questionable unknowns." In essence, the BDG used many of the same arguments for keeping the dam that supporters of removal made — fear for the watershed, recreational use of the area, and care for wildlife, as well as economic concerns.[62]

In their own letters and e-mails to the EPA and governor, dam removal opponents echoed the BDG's arguments for keeping the structure. Some, like Missoulian Daniel Monroe, reminisced about their recreational outings on the reservoir and the wildlife they encountered. In an hour of sea kayaking, Monroe recalled "spotting an eagle, blue heron, and a moose," plus other recreationists such as "canoeists . . . and . . . a double skull team." In his e-mail he made the point that most people who wanted the dam to remain also wanted the EPA to clean up the water, just not by draining the reservoir and getting rid of a viable power source. Don Wibracht wrote to the agency to point out that the dam "has created a habitat that nurtures a dense and diverse flora and fauna. Removal of the dam would result in the diminishment of a remarkably fecund wetland." Elden Inabnit of Missoula wrote to the *Missoulian* that a lifetime of living in the area had shown him that nobody was dying or suffering from toxins and that "the area above the dam on the Clark Fork looks like the 'Garden of Eden' every summer with boaters, moose, deer, birds and critters up the gazoo."[63]

Not all opponents of dam removal made arguments that mirrored those of the issue's supporters. Just as anticorporate sentiments motivated many proponents of taking out the dam and restoring the rivers, antienvironmentalism found its way into the rhetoric of many BDG supporters. Leroy Zent e-mailed Governor Martz from Great Falls to propose that if the EPA decided to remove the Milltown sediments, every environmental "activist must adopt and take home at least 100 lbs. of sediment and take care of it" and "dig into their own pocket and contribute the sum of $1,000.00 to help off-set the cost of this project." In an e-mail to the EPA, Gary Bray charged that he was "tired of all the damn dam liars. . . . Eviro [*sic*] groups the worst. They find one stinking fish on the river bank a fisherman threw there and they cry 'Fish Kill.'" His outrage over the situation ended with the supposition that "if telling lies were against the law there would be no Enviros. They would all be in jail!" Other antienvironmentalist sentiments were less accusatory. And in the end, there were far, far fewer opponents than supporters of dam removal, but they existed and included locals at the site of the dam.[64]

The Bonner Development Group tried to discredit the flood of support the EPA received for dam removal. In response to the EPA's receipt of thousands of letters favoring option 7A, Bruce Hall asked John Wardell, "What legitimacy does the United States Environmental Protection Agency place on the receiving of form post cards and petitions?" While disparaging the postcards that the CFC wrote, printed, and distributed as part of its "Remove the River. Restore the Dam" campaign, Hall ignored the fact that his group had mailed out 1,100 postcards to conduct its own survey months earlier.[65]

Methods aside, the EPA received more comments on Milltown than on any previous Superfund site. Most people credited the CFC campaign. As this public "yell" swelled into 2002, historian Stephen Ambrose donated a quarter-million dollars to the coalition and Trout Unlimited to help them further fund their dam removal campaigns. He argued that the Army Corps of Engineers "don't build many new dams these days, but are beginning to use their expertise to restore rivers and streams." By putting his money where the organizational mouthpiece was, Ambrose also signaled that he understood the importance of public education and public input when it came to this historic shift in managing watersheds.[66]

Such widespread public support for a thorough, permanent cleanup was critical to the Superfund process. That was certainly what the "Mother of Superfund," Lois Gibbs, thought when she came to Missoula to talk about the law, its history, and its future. Gibbs admitted that Superfund had struggled since she had helped birth it by fighting to get the federal government involved in making the Hooker Chemical Company clean up her hometown of Love Canal, New York. Yet, looking at the Milltown Dam issue, she reminded a crowd at the University of Montana, "When communities have been able to move it politically, Superfund has worked." Having followed the history of this Superfund site, she thought it was "headed for a solid, healthy cleanup" thanks to public advocacy. Like Gibbs, most commenters on the Milltown site filled their letters with more visions of renewal for the watershed.[67]

Following such leads, state support for a decision seemed to be heading in favor of dam removal. In January 2002, the *Missoulian* reported on a meeting of state officials, including the attorney general, wherein "every elected official in the room—Democrat and Republican . . . agreed. . . . The U.S. Environmental Protection Agency should order the removal of Milltown Dam." While reiterating all the familiar reasons for removal, such as recreation, economics, fish, and safety, the state attorney general, Mike McGrath, assured those at the meeting and the public that he would help make sure the MPC's sale of the dam to NorthWestern Energy and its parent company, telecommunications giant Touch America, would not leave the city or state government liable for the structure's fate.[68]

By March 2002, Republican governor Judy Martz's office hinted for the first time that the state was leaning toward dam removal. In a letter to the Rocky Mountain chapter of Montanans for Multiple Use, a group that supported state over federal authority in general and favored keeping the dam in particular, Martz countered the economic arguments for keeping the structure with the conclusion that "the Milltown Dam is currently an encumbrance for any corporation, agency, or group trying to generate electricity and revenue." The governor also forecast the state's leanings by writing about how removal would succeed. As if the decision were already made, Martz wrote that during dam removal "releases will be minimal and the Missoula aquifer will not be contaminated." The EPA effectively admitted a similar bias by the end of the month when John Wardell announced that his agency was going to begin sharpening "the cost estimates" for dam removal, according the *Missoulian*. If that wasn't definitive enough, Wardell admitted that "from a personal perspective the removal option is where I am leaning." By that time Wardell had nearly ten thousand postcards and personal letters that weighed heavily in favor of option 7A.[69]

The support of the Confederated Salish and Kootenai Tribes pushed the state toward dam removal as well. Because the tribes were around before the state and because the 1855 Treaty of Hellgate granted the tribes rights to traditional hunting and gathering grounds, including the "place of the big bull trout," tribal chairman Fred Matt believed that "the tribes have 'a little more' right to—and responsibility for—the Clark Fork than does the state of Montana." While maintaining a focus on the importance of bull trout recovery and habitat restoration, the tribes added the renewal of a spiritual resource to their reasons for supporting dam removal. Having gained official position as a "trustee" for the site by the Department of the Interior in 1995, the CSKT's claim to rights and responsibilities carried weight within Superfund law. But the tribal government was participating in the implementation of public policy beyond the confines of the reservation for more than just tribal reasons. In a letter to Governor Martz, Matt articulated the CSKT's support for dam removal as a way of preserving a subsistence tradition, guaranteeing treaty rights, and allowing "Montana citizens" to seek "to preserve the right under the Montana Constitution to a clean and healthful environment." The tribes were also acting within their Superfund rights.[70]

Aside from "public participation," Superfund law contains only one section dedicated to a specific group. Under section 9626, the law gives "Indian tribes" the "same treatment as a State with respect to the provisions" of Superfund. In particular, the law includes tribal "consultation" in the choice of a cleanup action. At Milltown, the CSKT exercised that right over an environment that was not on its reservation and over an issue that was of public, not just tribal, interest.[71]

In July 2002, the EPA added to the momentum of turning Superfund cleanup into environmental and economic renewal. In a news release entitled "EPA Announces $1.3 Million in Superfund Funding to Return Contaminated Sites to Productive Use," administrator Christie Whitman confirmed her agency's new commitment to turning some of the nation's worst waste sites into healthy, financially productive places. Milltown was one of nineteen communities being offered funding.[72]

The $40,000 Milltown was to receive as part of the new Superfund Redevelopment Pilot Program seemed to strengthen the possibility of dam removal. Nationally, the initiative indicated that Superfund was evolving into more than just a hazardous waste containment or cleanup policy. The Pilot Program would award money from a viable PRP or the EPA, if no PRP existed, to help local governments and community organizations plan and carry out post-cleanup ecological restoration and economic redevelopment. Funding for such efforts, which Milltown received at the end of July 2002, would become a standard part of Superfund implementation nationally. The program enabled communities to turn toxic waste sites into everything from "wildlife sanctuaries" to "manufacturing facilities."[73]

At Milltown, the pilot money helped give Bonner and Milltown residents more direct participation in the Superfund process and outcome. Although the Bonner Development Group opposed the decision to award the money to the county, even though that was what the program was legally bound to do, members of the group joined the project. In fact, it was the BDG's attention to how dam removal would undermine the community's tax base that kept local economic interests in the spotlight of restoration and redevelopment. The influence of the BDG was one more way in which Superfund implementation was a coordination of many competing interests rather than a winner-take-all outcome. The redevelopment committee that formed to direct the pilot project indicated that dam removal and river restoration was the favored solution at Milltown.[74]

More than just opinions and planning pointed in the direction of dam removal. By summer's end, the EPA had ordered a slow drawdown of the Milltown Reservoir to allow federal and state agents to test the toxic sediment. These tests were meant to help "refine . . . cost estimates of several cleanup options," although testing sediment so that it could be left in place behind a rubber-fortified dam seemed unlikely. Keith Large, an EPA site manager, told the *Missoulian* that the drawdown would also answer questions that spoke to removal, such as: "What construction techniques could we use to load sediments onto rail cars? How much excavation could we do versus dredging?" The month-long drawdown also allowed Montana FWP biologists to kill as many of the reservoir's illegally introduced, trout-eating, nonnative northern pike as they could net, hook, or strand on dry land.[75]

By mid-August 2002, the EPA had also eliminated another argument against dam removal. The agency released its plan for cleaning up the Clark Fork between Butte and Milltown. Funded by ARCO, the $100 million effort included removing the worst of the streamside contaminants. Many of the people who opposed dam removal argued that the dam prevented upriver contaminants from polluting the Clark Fork below Milltown. The plan to remediate and restore the upper river seriously weakened that argument. This "turning point," as Tracy Stone-Manning called the plan for the upriver portion of the nation's largest Superfund complex, was a nudge in the direction of restoration for the reservoir portion.[76]

Public comments kept pushing the EPA in that direction as well. By early fall 2002, the EPA had counted more than nine thousand public comments on Milltown. A growing number of those comments came from Bonner-Milltown area residents who now favored dam removal. Most of those writers spoke of a silent local majority being out "yelled" by the BDG. As Mary Erickson offered, "we are part of a large but perhaps less vocal or financially motivated group of citizens that want to see the whole 'dam' mess cleaned up." She admitted having canceled her membership with the BDG because of the influence she thought ARCO was having with the group.

By October, Erickson and many other locals had raised their voices by forming the Friends of Two Rivers group to counter the BDG in the Bonner-Milltown community. On the eve of Halloween, the new group's efforts were part of more than ten thousand comments received by the EPA, prompting Diana Hammer, who directed the site's public relations, to say that it was "very rare, unheard of actually, to get this kind of response outside of a formal public comment period."[77]

The EPA announced that it would release its cleanup proposal for Milltown "in late January/February 2003," although it still awaited "State and Community Acceptance." The subsequent comment period was a Superfund formality. The EPA already knew what the public thought. As Stone-Manning said about one aspect of the CFC campaign, "we did sort of a pre-comment period campaign so that they [EPA] knew they would have public support." The governor, however, had opted for public silence.[78]

The CFC and other dam removal advocates fretted about the governor's position. Referring to George Bush's campaign promise, Stone-Manning recognized that Martz was "a Republican Governor under President Bush . . . who on the campaign trail said there would be no dams coming down under his watch." In fact, after Bush had defeated Al Gore in 2000 and Martz had won the Montana governorship, the Milltown project manager had called Stone-Manning and asked if her organization was going to continue its campaign. She told Forba and the *Missoulian* that the national and state turn in politics was simply motivation to "ramp up" the campaign. The plan was to trump Republican opposition to dam removal at the state level with public support. The CFC didn't foresee turning such opponents into supporters, or the ultimate lack of partisanship on the issue.[79]

Addressing dam removal advocates and opponents, Martz was beginning to intimate her decision. Responding to concerns expressed by the BDG about the local economy, the governor detailed the high cost of stabilizing the Milltown Dam and its poor prospects for turning a profit. She admitted working closely with other agencies on the issue, such as the state FWP, USFWS, and the CSKT, all of which supported option 7A.

In her State of the State address on January 21, 2003, Martz made the surprising announcement that she and the state would support dam and sediment removal. She said that safety issues, which had gained attention when FERC had classified the dam as a possible post-9/11 terrorist target, had swayed her. The recent organization of the Friends of Two Rivers had also helped Martz recognize that there was ample local support for option 7A. The next day the *Missoulian* reported that the EPA had come to the same conclusion. In February the agency would make public a "$90 million proposal" that specified "reconfiguring the Clark Fork River's channel, restoration of the confluence of the Clark Fork and Blackfoot rivers, and a slurry pipeline that delivers contaminated sediments to a permanent repository on high ground south of the river." Even President Bush made an exception to his blanket support of dams by giving the Milltown Dam removal a nod.[80]

In the three years between the beginning of the CFC campaign and Martz's public embrace of dam removal, Tracy Stone-Manning had had a dream about the confluence of the Clark Fork and Blackfoot. Admitting that it sounded "goofy" for someone who was not "New Agey," she recalled falling asleep and simply hearing the sound of the free-flowing rivers converging. When she woke she told her husband, "It is going to happen." When Stone-Manning heard Martz's address, in which the governor "sounded like she was reading off our website," the coalition uncorked champagne. Listening to Martz reiterate the rhetoric of the campaign confirmed the message of Stone-Manning's dream confluence. It was indeed "going to happen."[81]

More support for making it happen quickly followed. The state's congressional representatives made visits to the dam to publicize their position in favor of its removal. Martz revealed that the state had been in secret negotiations with ARCO, the dam's owner, and the EPA, which pointed to the companies' acquiescence to the proposed plan. "They could have fought us and paid for attorneys from here to forever," said Martz of the liable parties' willingness to accept the EPA decision. She knew that did not eliminate the possibility of problems along the way to a cleanup plan. In a letter to the Missoula county commissioners, she wrote, "Differences of opinion will occur and tough decisions will need to be made." But the path seemed set.[82]

The buyout of ARCO by BP in 2000, which frightened many people concerned about Milltown, helped facilitate dam removal in the end. State director of Trout Unlimited Bruce Farling recalled meeting a small group of BP scientists and executives for the first time during the feasibility study process. He and Tracy Stone-Manning explained their case for dam removal to the BP representatives, who, as Farling intimated, thought "we were like these loony, crazy people." With ARCO environmental remediation director Sandra Stash looking on and defending the dam-in-place logic, Farling saw the BP "guys" experience a sudden "attitude shift." They realized that Milltown was, according to Farling, "this annoying pimple thing," and they were a "multi-zillion dollar company." Stone-Manning saw the shift, too. As opposed to what Stash had been telling them, the BP "guys" suddenly understood that removing the dam would erase a "tiny liability . . . off our books." Getting the Superfund site off BP's books permanently also coincided with the EPA's increasing preference for permanent remediation.[83]

One of the most contested elements of the plan was where to put the contaminated sediment. That struggle was part of a new debate. How to remove the dam and restore the river rather than whether to do so became the central question at Milltown.[84]

With a Record of Decision designating the final remedy due out within the year, the EPA fostered public comment on its plan more than at any other time during the process. The comments came in a steady trickle compared to the surge that preceded the agency's decision. Most writers praised the option to remove the dam as a good, foregone conclusion and directed the EPA to supervise a thorough cleanup and restoration. Acknowledging both the precedent Milltown set in the public participation schedule and the need for that engagement to continue, John Wardell told the *Missoulian*, "This effort has been unique because we've gotten so much public comment in advance of the release of the proposed plan. But we need to hear from people now as well." Diana Hammer later acknowledged that Milltown became a model for how her agency engaged the public in the Superfund process, such as commencing official comment periods earlier.[85]

Hammer also recognized that the EPA was expanding its definition of Superfund cleanup. "When the Superfund program first started, we were dealing with . . . oozing messes or leaking barrels and drums," she recalled. With Milltown as an early and successful example, Hammer acknowledged that her agency "should be thinking about how that land could be re-used in some beneficial way." Until the EPA started the Redevelopment Pilot Program and places like Milltown looked to a usable, economically viable future, simply containing wastes "just didn't seem like we had finished the job" to Hammer.[86]

As Milltown moved toward the goal of restoration, so did Superfund in general. Even before the Pilot Program, CERCLA had enabled Natural Resource Damage claims by federal, state, or tribal trustees of publicly held environments, such as many waterways. Trustees could sue PRPs for damages to restore those environments beyond what Superfund cleanup entailed. Montana's NRDP settlement from ARCO provided $28.9 million to restore Milltown.[87] After dam removal, cleanup, and restoration commenced at Milltown, the EPA released a watershed restoration guide for federal and state project managers. CERCLA topped the list of federal programs for which watershed restoration was a concern. The agency's guide featured Milltown as one of its case studies.[88] Eventually, the EPA began incorporating restoration as a de facto goal of Superfund cleanup. By 2010, the agency was studying ways to reduce the carbon footprint of Superfund cleanups and to make remediation work generally more "green." Greening cleanups was in addition to "satisfying threshold requirements" of Superfund remediation "intended to restore contaminated lands." The EPA's baseline for CERCLA actions had shifted, in its own definition, from containment of wastes to restoration efforts.[89]

The path to dam and sediment removal at Milltown furthered other changes in environmentalism as well. The Clark Fork Coalition's campaign spoke of a new mode of environmentalism. The group created surprising images of environmentalism as fun, funny, and even "sexy," as Tracy Stone-Manning put it. Such playfulness in a media campaign advocating an environmental issue allowed the group to entice, educate, and motivate people outside its usual audience. Those people participated in the Superfund process, altering its course in historically large numbers. E-mails were quickly taking the place of traditional letters. In addition to the tone and style of the coalition's campaign, it promoted the idea of pushing Superfund remediation into the realm of restoration. The motto "Remove the Dam. Restore the River" connected Superfund's original mandate of basic waste cleanup with the new direction of ecological regeneration. And the emphasis that opponents, such as the Bonner Development Group, placed on the local economy as a Superfund consideration coincided with the EPA's fledgling efforts to promote redevelopment along with remediation. As the agency's public information arm, the Clark Fork River Technical Assistance Committee, deemed it, the "project now underway at the Milltown Reservoir is one of the nation's most challenging and ambitious environmental cleanups" because of its size as well as its integration of restoration and economic development with cleanup. At Milltown and elsewhere, this would come to be known as the "three Rs" of Superfund: Remediation, Restoration, and Redevelopment.[90]

The drinking of champagne in celebration of the EPA's decision at Milltown hardly ended the Superfund process. A week after its announcement, a *Missoulian* editorial cheering the proposal concluded that "a decade from now, when the dam is gone and the river is cleaner, people will look back in amazement that we ever agonized over what course of action to take with Milltown Reservoir." In the following ten years, the dam came down to many rounds of toasting and festivities. Much of the agonizing abated. The river began to regain some of the character that earned it the name "Place of the Shining Waters" in the Salish tongue. The EPA, state of Montana, Confederated Salish and Kootenai Tribes, and local government helped promote new economic opportunities associated with the free-flowing confluence. Restoration and redevelopment also spread far beyond the place where the rivers meet and the dam once stood, and it did so in very unexpected ways. The whole process settled into another chapter in the history of the area and of Superfund. As it did, two of the truly unique features of the final stage of the Superfund process at Milltown—the actual work of the three Rs—were the rise of corporate ecological restoration and the directions people pushed that restoration.[91]

CHAPTER 7

The Three Rs of Superfund

Remediation, Restoration, and Redevelopment

Restoration is about having the power to visualize, to say that we can imagine a landscape that we don't see today, that we can create, or recreate, a landscape that was seen by Lewis and Clark. . . . We can look to the past, and by understanding the past, visualize the future. And then engage communities and conservationists in the act of restoration. That has a lot of magic and power.
— Bruce Babbitt on the theme of his time as Secretary of the Interior

Spring runoff from the Clark Fork had threatened repeatedly to breach the Milltown Dam in the century it stood. The history of floods imperiling the dam helped justify its breaching. But heavy machinery, not high water, did the actual work.

On March 28, 2008, a few hundred people stood on a bluff overlooking the dam site. With high noon sunshine melting the snowpack off the surrounding mountainsides, Montana governor Brian Schweitzer hollered, "Let 'er run!" A stream of muddy river water flowed in a new channel past what for a century had been the north end of the Milltown Dam. Construction workers, public officials, environmental advocates, and local citizens all cheered. From the bluff above the confluence's south side, people spent the next few hours watching a trickle swell into a modest waterfall that tumbled over the twenty-foot drop where the dam used to be. They snapped photographs, listened to short speeches, and pointed out curves in the river's new channel as well as remnants of its creation such as berms, piles of riprap, orange pylons and plastic fencing, earthmoving machinery, and a nearly two-hundred-acre expanse of muddy ground that had been underwater for a hundred years. County Commissioner Barbara Evans offered that it was "really a historic occasion."

If dynamite and a massive wave of water played no part in the river's first uninhibited run past the dam in a century, no one seemed let down: Speakers reminded the crowd that after thirty years of Superfund designation, the celebration was about the culmination of a long, tedious, sometimes contentious, but very methodical process. It was also about the future. The first breaching of the dam was one of many riverside ceremonies

Preparatory to sediment and dam removal, reservoir water was routed into the channel on the right (November 2007), and the sediments, surrounded by the light-colored rock berm, were prepared for removal. Rail cars to carry the wastes sit poised on the tracks between the removal area and the channel. Before the dam would be breached, sediments would be removed from the area on the left near the interstate and a new channel would be constructed, where the river would then be diverted. *Courtesy Diana Hammer, photographer, U.S. Environmental Protection Agency, Helena, Montana.*

Water flowed through the former Milltown Dam on March 28, 2008, after the dam had been breached, starting as the minor gush of water seen here. *Courtesy Diana Hammer, photographer, U.S. Environmental Protection Agency, Helena, Montana.*

commemorating the many small steps that marked fixing one of the nation's largest Superfund sites. It would take at least another year just to get rid of the rest of the dam so that the Clark Fork could truly run free.

The same kind of planning, bureaucratic oversight, corporate negotiations, and public input that brought about the decision to remove the Milltown Dam and its load of toxic sediment prevailed in the years the actual dam removal took. Numerous celebrations marked milestones of both the paper and shovel work of Superfund remediation at Milltown. School groups toured the dam's powerhouse before it closed for good; workers cheered the first bucket of sediment scooped from the muddy floodplain; and on any given day, people trekked to the site's overlook to watch the earthmoving, jackhammering, or carving of a new river channel. Most days onlookers would have had a hard time distinguishing cleanup from

This view of the removal site shows the river channel on the left. At this point, all the sediments are gone except in the area under the railroad tracks. In the foreground, the dam's spillway is being removed. *Courtesy Diana Hammer, photographer, U.S. Environmental Protection Agency, Helena, Montana.*

Restorers used logs and boulders to stabilize the banks during the rechannel-ing of the Clark Fork and planted locally sourced willows, alder, cottonwood, and other native plants to re-create viable riparian fish and wildlife habitat and make the area safe for people to enjoy. The restoration area extends more than three miles upstream and covers more than five thousand acres of new flood-plain. *Courtesy Diana Hammer, photographer, U.S. Environmental Protection Agency, Helena, Montana.*

river retooling, much less the goal of making the project an economic boon to the area. Yet the integration of remediation, restoration, and rede-velopment made Milltown "really" historic, as Evans put it.

As EPA community coordinator for Milltown, Diana Hammer saw how the process of implementing a Superfund solution at the site helped make it a model of the law's evolving emphasis on "the three Rs." As she awaited Schweitzer's command, Clark Fork Coalition scientist Christine Brick told the *Missoulian*, "I feel like an 8-year old kid waiting for Christmas." The Christmas present was $120 million for remediation. But with over-whelming local support, cleaning up the toxins at Milltown became the starting point for restoring the watershed. Part of that evolution was figur-ing out what restoration of a natural environment meant. As Brick once observed, "without restoration the river would have been left to run in a rock-lined ditch."[1]

Environmental historians have argued that restoration in the West began with efforts to return individual species to their native places. In the late twentieth century wolves returned to Montana, for example. In *The Natural West*, Dan Flores proposed that the bigger challenge was for westerners to tackle whole ecosystems, such as watersheds, and to do so with the understanding that virtually all environments are products of a long and varied human presence. Thus, restoration would be another new human manipulation of the environment, not a return to untrammeled nature.

No one involved at Milltown suggested that Superfund implementation would return the area to pre-European conditions. Overlooking the site of the former dam and toasting its removal included watching cars and trucks roar by on Interstate 90 directly opposite the confluence. Remembering that the area once bore the names "place of the big bull trout" and "shining waters" was more about recognizing that people could help undo some of the last hundred years' worth of industrial damage to the environment, rather than hoping that they could wholly erase the world wreaked by industry. Milltown's restoration in the history of Superfund also demonstrated that that new environmental project was a national undertaking, not just a western one. As Elizabeth Grossman pointed out in *Watershed: The Undamming of America*, cases like Milltown were part of a national trend whereby more dams have been decommissioned than built in the United States since 1998.[2]

At Milltown, restoration came to include dispersing Superfund money and addressing impacts well beyond the confines of the site. Corporate and public interests, federal and state laws, Native American efforts, and the riparian environment all played a part in determining the course of dam removal and river restoration at Milltown. So, too, thinking about how to repair and revitalize the river led to considering the health of the local economy. Superfund implementation aided a rise in the environmental restoration business and provided a catalyst to local entrepreneurship.

Superfund remediation meant removing or containing the source of toxins that threatened human health. At Milltown that effort centered on removing contaminated sediments. The debate over where to move tons of contaminated sludge brought with it the question of environmental justice. The choice to move Milltown's waste to an upstream community came about because of ecological concerns and local advocacy more than social disparity or corporate malevolence. Just as remediation affected environments outside the Superfund site, restoration spread beyond the designated site as well. One of the most unique dispersals of Superfund impacts happened on a tributary to the Clark Fork watershed on the Flathead Indian Reservation. Finally, Superfund implementation at Milltown demonstrated how improving an environment could help redevelop a fading extractive economy. As a 2011 EPA case study of Milltown's accomplishments stated, integration of the three Rs, backed by robust and consistent community involvement, was a model "that can help guide similar projects at contaminated lands across the country." Public participation in the Superfund process at Milltown became part of a new, more thorough definition of toxic waste cleanup for the EPA.[3]

~ ~ ~

Heavy metal–laden sediment behind Milltown Dam was the culprit that earned the site federal designation and it was the first major consideration when cleaning up the place became a reality. As Milltown project manager Russ Forba told *Waste News* shortly after the EPA released its first cleanup plan, "the main emphasis of this and the main cost is the removal of sediments to restore groundwater."[4] EPA estimates showed that "300,000 cubic yards of uncontaminated sediment scoured from the mouth of the Blackfoot River" would wash downstream in the months following dam removal. In the next ten years the Clark Fork would probably flush another ten times that much sediment, "the largest sediment load ever released by a dam removal in the United States," according to the federal agency's remediation plan. As monumental and iconic as dam removal was, it was the removal of the contaminated sediment that stirred the first crucial debates about how to implement the EPA's decision for Milltown.[5]

The public comment period that followed the EPA's draft plan elicited fervent response to the question of sediment removal and forced one of the major changes in the agency's plan. It also complicated historical arguments about the overlap between social justice and environmental issues. Environmental historians have argued that polluted environments disproportionately affect working-class and minority populations. Either those populations can't afford or don't have access to the political power and scientific knowledge to fight pollution, or the low property values surrounding polluted places draw disadvantaged populations. A caveat to that argument is that people who *can* afford the efforts maintain a "Not in My

Backyard" attitude, known as "NIMBYism," toward environmental pollution. At Milltown, the sediments wound up at a significant remove from
the Superfund site. The forces behind that decision challenged traditional
arguments about NIMBYism. Milltown had the same history and contemporary demographics as a working-class community as the place the
area's sediments ended up. The reason an environmental justice controversy arose over the movement of Milltown's contaminated waste was that
Missoula residents were so active in getting them moved. Their destina-
tion, Opportunity Ponds, Montana, certainly suffered from its proximity to
industrial waste. But injustice within the Milltown Superfund process did
not account for those circumstances. Mining history, public opinion, and,
most importantly in the case of Milltown, the know-how of a new corporate restoration industry shaped the sediment decision.[6]

The issue of where to put contaminated sediments gained momentum
as early as 2000. A study contracted by the EPA evaluating disposal sites
estimated that a site needed between 15 and 184 acres of land, depending on how much of the sediment the final cleanup produced. The study
identified two local sites. Using either meant creating a new, plastic-lined
disposal complex on undeveloped land on the margins of the Clark Fork
floodplain.[7]

Others had local waste repository sites in mind as well. In August 2000,
Peter Nielsen wrote to Russ Forba to tell him that the Missoula City-
County Health Department had identified no fewer than "eleven sites" in
the county that might work.[8] The state's Trout Unlimited director, Bruce
Farling, thought that most of the local environmental community agreed
that it "was fine having it [sediment] around here somewhere as long as it's
high and dry," as well as closely monitored. That opinion wasn't the same
as saying every environmentalist should take home a share of the waste,
as one opponent of dam removal had once suggested, but it wasn't a "Not
in My Backyard" response, either. It was a practical understanding that
cleaning the river meant sacrificing some other landscape. At the time, a
local place was acceptable.[9]

In the end, the practicalities changed. Local sites were too close to the
water. In April 2002, state EPA director John Wardell wrote to Dan Watts,
the president of Montana Rail Link (MRL), about the economics of using
his company to haul sediments away from Milltown. Dennis Washington,
who had grown his fortune in the 1980s by buying most of the mining in
Butte, owned MRL. Having seen ARCO saddle itself with environmental

liability, Washington purchased entities like Butte's East Continental Pit from ARCO, while leaving that company with all the liability for past contamination. As Butte-born journalist Edwin Dobb put it in an article for *Harper's Magazine*, "Washington's net worth . . . reportedly shot up to more than $1 billion, the bulk of it made in Butte on an investment of only $18 million, the fire-sale price ARCO accepted for the Richest Hill on Earth." Wardell's letter helped position Washington to make even more money from ARCO and its environmental liability. The state EPA director was opening discussions with Watts about getting "the best possible estimate for the potential cost of rail transport of the dewatered sediments from Milltown to Opportunity Ponds." With Watts's positive response, the possibility of moving the waste about a hundred miles upriver and spreading it in an existing repository near where it was produced became a reality. Watts roughed out that possibility at $20 million to $34 million depending on the amount of sediment, not including the construction of new rail spurs. With that, Opportunity Ponds became central to the debate about the Milltown solution.[10]

When the EPA released its proposed plan for Milltown a year later, on April 15, 2003, it still included placement of the sediments "in a lined repository located less than a mile downstream from the Dam." The plan was to use "hydraulic dredges" to scoop the contaminated sediments off the bottom of the reservoir and "send them via slurry pipeline" to the disposal site.[11] As with so many turns in the Milltown story, vigorous public participation prompted an extension of the EPA's official comment period on its dam removal proposal.[12] In a newspaper article about the comment period, project manager Russ Forba encouraged people to help answer questions about the how and where of sediment removal. Opportunity Ponds became the favored location.[13]

In 1914 ARCO's predecessor, the Anaconda Company, established what a *New York Times* retrospective called "a model community to show that people could raise crops and livestock" in the midst of heavy mining and known pollution. From its inception, Opportunity Ponds had been a dumping ground for toxic waste. It had streetcar service, a school, and a golf course, and people had readily bought ten-acre parcels and moved their families to the new town of Opportunity as it boomed alongside copper production during and between both World Wars. Residents accommodated the waste repository as a normal, necessary part of life in the heart of industrial mining country. But with most mining jobs gone by the early twenty-first century, many locals began to balk at having to suffer the industry's consequences.[14]

Forba, along with the EPA's John Wardell and Diana Hammer, received plenty of answers to the question of where to dump Milltown's wastes, especially during public forums on May 7 and 8, 2003, in Bonner and Missoula. The most common comments about the agency's proposal addressed moving sediments to Opportunity. Local resident and Friends of Two Rivers member Mary Erickson opened the discussion by asking, "Why purchase another property here, prepare it, line it, build the site and duplicate that kind of expenditure when perhaps we could utilize the Opportunity Ponds?" Locals and people from up and down the river agreed that using undeveloped land on the lip of the river's floodplain for a toxic waste repository that had to last in perpetuity was a questionable notion of permanent cleanup. Proposals ranged from stuffing the wastes back into the Butte mines and pits from which they came, to the more practical and thoroughly supported option of the Opportunity Ponds.[15]

The day before the Bonner and Missoula meetings commenced, U.S. senator Max Baucus weighed in on Milltown. In a letter to Russ Forba, Baucus summarized how an Opportunity repository "would allow EPA to coordinate and manage in one place" all the waste produced from cleaning up the entire Clark Fork Superfund complex, with plenty of space available if that amounted to more than expected. Along with that justification, Baucus offered to help negotiate a deal between the "interested parties," which meant getting ARCO to pay Dennis Washington's enterprise for the work of dealing with the kind of waste that Washington had been savvy enough to avoid liability for in his purchase of Butte mines from ARCO. Baucus added that his proposal would create more Montana jobs as well. The Missoula-based Carpenters Local Union No. 28 and the Montana Community-Labor Alliance/Jobs with Justice registered its support of the dam removal proposal and favored the Opportunity Ponds option as a jobs creation or redevelopment prospect, too.

State senator Sherm Anderson, who represented the district encompassing Opportunity Ponds, denounced Baucus's proposal to the EPA, claiming that the area had already suffered enough from being a waste repository and that moving Milltown's sediment the one hundred miles there would cost an additional $100 million. Besides grossly exaggerating the added cost of freighting the material upriver, compared to building a new site near Milltown, Anderson misconstrued the fact that the addition to Opportunity would be a fraction of what was already there. In his recent book investigating the story of moving sediments upriver, *Opportunity, Montana: Big Copper, Bad Water, and the Burial of an American Landscape*, Brad Tyer quantified Milltown sediments by recounting the removal process whereby "every morning for . . . two years, Montana Rail Link left forty-five empty cars on the loading siding. Every evening, Envirocon left forty-five cars filled with a hundred tons per car of sediments infused with heavy metals. Every night for two years, Montana Rail Link ran the trains to Opportunity, where Envirocon excavators transferred the dirt to dump trucks, drove them to an eight-hundred-acre patch of the Opportunity Ponds, and spread them two feet thick on the ground."[16] From the perspective of the extant waste repository, the EPA estimated Milltown's waste would need less than 180 acres, whereas the Opportunity site was 4,000 acres.[17]

Ecological arguments also pushed the EPA to change its plan for Milltown's sediments. A Missoula ecological consulting firm sent a four-page letter to Russ Forba outlining its stance on the Milltown proposal. The letter argued that moving sediments to Opportunity would help make the "fullest possible re-naturalization of the river/floodplain/riparian area" possible. Penning the Missoula City-County Health Department's eight-page letter to the EPA, Peter Nielsen said that the first priority was getting the sediment out of the river and if that entailed creating a local repository, local government would support it. But, he confirmed that it made more ecological sense to use the Opportunity site instead of breaking "clean ground."[18]

Corporate opinion agreed. In early 2003, Missoula-based Envirocon, with ARCO's encouragement, had won the bid for performing sediment cleanup at Milltown. In 1988 Dennis Washington had started Envirocon to apply the "heavy construction" resources and know-how of major industry, like mining, to environmental cleanup. Having performed such work on thousands of projects, many of which fell under the direction of federal agencies, the company helped shape the Milltown sediment removal plan as well as carry it out. At Envirocon's first meeting with local government and the public, the company's regional manager, Kris Kok, explained the technical and ecological wisdom of draining the reservoir and excavating dry sediments with backhoes and dump trucks instead of scooping and sucking up wet muck with floating machinery. For one, Envirocon could do the job in half the time since crews could work longer days and year-round on dry ground, as opposed to discontinuing work most of the winter because of a frozen reservoir if the pool were left in place. Even at two and a half instead of five years, the project would be Envirocon's largest, its most lucrative, and its first in Missoula County. Kok also proposed that the EPA retool its plan by rail freighting the contaminated soil to Opportunity. Acknowledging that his company had no real say in the matter, Kok echoed the suggestion that using the Opportunity site would eliminate building and monitoring another repository.[19]

Opportunity residents were less excited about the possibility of acquiring Milltown's mess than Milltown residents were about sending it upstream. When Kok explained his company's preferred plan to people in Anaconda–Deer Lodge County, they had more concerns than praise. They wondered why the sediments were called toxic waste in Milltown but described upriver as relatively good organic material that would help "revegetate" the Opportunity Ponds repository. Area residents believed it was because of the community's declining population, high unemployment, and poor wages compared to the relatively opposite situation in Missoula. Kok and ARCO representatives explained that location mattered. When in the river and Milltown groundwater, the stuff was a health concern worthy of Superfund designation; as a covering for wastes with much higher concentrations of heavy metal, Milltown's sediment would be an enhancement. Still, some opposition remained, although it seemed infused with an anti-Missoula sentiment as much as it seemed concerned about adding a bit more material to Opportunity's repository. Anaconda resident Charlene Hagan proposed turning Opportunity's "lose-lose situation to a win-win" for Missoula. In her letter to Forba she asked whether the sediment could be "unloaded and sculptured, then capped at a proper site . . . to construct a world class ski jump hill (or hills) for different height jumps as well as acrobatic jumping." Hagan's assessment of Missoula as a town of ski enthusiasts was more about aiming sarcasm downriver than proposing a viable waste containment site. And, as Bruce Farling put it, he and Peter Nielsen had traveled to Anaconda and Opportunity over the years and gotten figuratively "beaten up" for trying to help residents push ARCO to improve its sprawling waste repository.[20]

While Opportunity area residents were slow to accept the new proposal, others did so quickly. Governor Martz wrote to John Wardell supporting the change in the EPA's plan. By May 2004, the EPA had released a Revised Proposed Cleanup Plan, which acknowledged that along with public input, ARCO's promotion of the Envirocon method of removing sediment had swayed the agency. The new plan spelled out its "key revisions" as draining the reservoir and rerouting the Clark Fork temporarily to help dry the sediment, removing 2.5 million cubic yards of it mechanically (instead of using hydraulic methods), and transporting the waste by Montana Rail Link to Opportunity Ponds. Although the revisions inflated the estimated price tag of cleanup from $95 to $106 million, the EPA clarified that public opinion and the responsible party were helping drive that change. At the same time, the EPA released its Record of Decision for the upper Clark Fork, which included sending toxic sediment from nearly 100 miles of riverbank to Opportunity Ponds. All the sediment being removed from 120 miles of the Clark Fork Superfund complex, not just Milltown's relatively modest portion, was destined for Opportunity. That helped squelch the reality of building separate site-specific repositories throughout the floodplain. So did a steady flow of public input.[21]

Along with Milltown and Missoula area residents, some people from upriver, like Anaconda–Deer Lodge County resident Jim Flynn, wrote to the EPA voicing the opinion that "removal of the Milltown Tailings to the Opportunity Ponds is also a good and positive step forward." A Butte "extractive resources engineers and geosciences" firm produced a technical report confirming Envirocon's assertion that dry excavation of sediments would reduce the amount sent inadvertently downstream. A coalition of seven Montana NGOs, ranging from labor alliances to the Clark Fork Coalition, coauthored a letter to Russ Forba praising his agency's revised plan. Scores of citizens provided their own postage or sent e-mails to the EPA offering support for its new plan. By mid-June 2004, even Dan Cox, the chief executive of Anaconda–Deer Lodge County, informed Forba that his office favored the decision to dump Milltown's sediment at Opportunity Ponds, although he insisted that the effort receive as much attention to protect human health as the dam site.[22]

Milltown's cleanup added 2.6 million to the nearly 200 million cubic yards of mining waste already stored in the ARCO-owned waste facility. Instead of trying to find the more than $2 billion it would take to clean up Opportunity Ponds by hauling that waste somewhere else, the EPA required ARCO to monitor local water wells and airborne dust, which was the primary complaint of local residents. The company also maintained a trust fund to pay for future problems should they arise. ARCO and the EPA both supported adding Milltown's sediment, in part because studies showed that it would help abate dust by capping the existing contaminants with a layer of viable topsoil. Still, opponents described the issue in class terms—white-collar Missoula was pushing Milltown's contaminants on blue-collar Opportunity. It looked like classic environmental injustice. In reality, the decision had far less to do with polarized communities than ecological considerations. As much as anything, the solution to Milltown's sediment illustrated a major element of environmental cleanup and restoration: mending one place meant sacrificing another. Contrary to a few letters the EPA received, it wasn't possible to stuff the sediment back down mining shafts. It was possible to consolidate it in a single, long-extant repository.[23]

When finally the EPA favored the upriver storage of Milltown waste, it was not choosing a disenfranchised community for dumping over an affluent one piqued with NIMBYism. Nor was it opting for a cheap solution to handling pollution. The agency was picking an established, large, and safer place to dispose of the waste instead of creating a new repository on the margins of the floodplain. Because of transportation costs, using the established site *raised* the cost of cleanup. And, both communities in question shared an industrial, working-class history, just as both communities garnered EPA attention throughout the process. In addition, opposition in Opportunity led to closer regulation of its repository. Finally, the trainloads of waste that began leaving Milltown in early October 2007 ended up hauling 2.6 million cubic yards of sediment upriver, leaving nearly double that amount in the Clark Fork floodplain where the river had carried it and the Milltown Dam sequestered it over the last century. The majority of the sediment stayed put and became part of the restoration effort at Milltown.[24]

Using Milltown's sediment as "topsoil" for the Opportunity Ponds turned out to be more difficult and expensive than originally planned. By 2010, the EPA recognized that the Milltown sediment was failing to support the vegetative growth that was supposed to cap the ponds. Subsequent testing in a greenhouse showed that the sediment had more metal and less organic content than it did when it was underwater at Milltown. By early 2012, ARCO was committed to spending another $15 million to treat the Milltown sediment with lime and then cover it with an additional foot of clean soil, which on-site testing showed should support the necessary vegetative growth to complete remediation at the ponds.[25]

~ ~ ~

Other controversies besides the sediment issue arose in the process of implementing the Superfund cleanup at Milltown. The EPA and state of Montana had to file court proceedings against NorthWestern Corporation, the dam's owner, and ARCO to prove that the private agreement the two companies made limiting NorthWestern's liability lacked legal teeth. Champagne and celebration greeted the EPA's release of its Consent Decree in August 2005. The decree obligated NorthWestern Corporation to surrender its dam license, pay $11.4 million toward removing the structure, and give the state its land and water rights in the Milltown area, as well as a few miles of riverside property along a downstream section of the Clark Fork renowned as a whitewater canyon. As for ARCO, it had to pay $100 million for cleaning up Milltown. From locals to federal officials, people gathered on the dam's overlook to toast the decree, which also committed the state to spend $7.6 million (from its previous suit against ARCO) on restoration work at Milltown.[26]

With the first steps toward dam removal scheduled for as early as January 2006 and the money to do the work secured, the EPA had the intervening time to press forward with a few important legal provisions of the Superfund law. The agency had released its Record of Decision on the last day of spring 2004. Although the document was the final decision to remove Milltown Dam and sediments, the specifics of the work took time to determine. Those specifics varied from engineers figuring out the realities of dismantling the dam and powerhouse, to fisheries biologists preparing studies on how the process would alter fish survival and populations, to an Army Corp of Engineers ice researcher determining how the dam's absence might change the formation and dynamics of ice floes on the Clark Fork. The planning and scrutiny would last through 2006, when actual remediation work would commence.[27]

Even with the dam still standing, most of the public attention shifted to the prospect of restoring the Clark Fork and Blackfoot. Just as the public had absorbed other campaign language such as the "toxic time bomb," people followed the CFC's lead in asking the EPA to "dovetail" Superfund cleanup with restoration. The public, the CFC, and eventually local government and the state encouraged the EPA to make sure that money and time spent dealing with sediment and dam removal would make restoring a natural, free-flowing stream more efficient. Because of its responsibility for recovering bull trout, the U.S. Department of the Interior pressed the EPA to make restoration an "integrated" part of remediation, too.[28]

Media attention on the Milltown Dam Superfund site tended to emphasize visions of its restoration. In the years between the EPA's Record of Decision and Governor Brian Schweitzer's call to "let 'er run," newspaper articles about the project included a diagram depicting the confluence of free-flowing rivers. When the *New York Times* published a story about the dam's removal, the quip it included from Tracy Stone-Manning celebrated "putting a river back together" rather than taking apart a dam. Plenty of dams had come down in America. The removal of the Edwards Dam from the Kennebec River in Maine became a model for such projects in 1999. According to the nation's leading river conservation organization, American Rivers, "the number of recorded dam removals has grown each year" in the decade since the Edwards came down. Dismantling Milltown's dam would help revitalize the Clark Fork and the river made famous by Norman MacLean and Robert Redford, going beyond just removing the dam, and doing so as the first such project at a Superfund site.[29]

Lots of attention went to figuring out what restoring the confluence meant. Studying rocks, soil, and landforms to re-create the Clark Fork's historic riverbed, its natural drop, the particulars of its union with the Blackfoot, and its inclinations below the dam site had to be modeled and undertaken. Site-specific restoration also included revegetating the floodplain with native plants. Natural processes would account for some of the restoration. In one surprising and feel-good twist to the process, many native plants came out of a hundred years of dormancy under the reservoir and sprouted on their own in its absence.[30] The EPA estimated that it would take four to ten years for the Milltown aquifer to clean itself and become a viable water source for residents again. Meanwhile, Envirocon hauled sediment from riverbed to dump truck to train car, sculpted bends and cut banks, entangled logjams, and graded the floodplain around the confluence. Lots of that restorative work happened in the course of removing contaminated sediments but took place before ARCO had agreed to pay the state's Natural Resource Damage Program. Eventually ARCO's NRDP responsibility totaled $11 million for restoring Milltown, on top of the $168 million it owed for the same purpose throughout the watershed. In the end, restoration at Milltown was practical, scientific, cost-conscious, and driven by the desire to make the Clark Fork a free-flowing river and a healthier ecosystem. As the EPA-funded Clark Fork Technical Assistance Committee defined it, restoration was, at its simplest, "returning natural functions" to the river, which meant re-creating viable riparian fish and wildlife habitat.[31]

The EPA's oversight of restoration at Milltown also allowed that restoration wasn't always site specific. Some work financed by ARCO's restoration dollars happened well beyond the boundaries of Milltown or the Clark Fork's main body and stretched the definition of restoration beyond returning a waterway to its natural state. The Confederated Salish and Kootenai Tribes on the Flathead Indian Reservation undertook one of the most unexpected of those radiations, far from the confluence, known in Salish terms as the "place of the big bull trout." Tribal efforts on behalf of the threatened bull trout were part of a cultural restoration project as well.

Using funds garnered in 1998 from participating in the state's Natural Resource Damage Program, the CSKT chose to start a Bull Trout Restoration Project on the Flathead Reservation, about fifteen miles north of Missoula. Germaine White, director of the project, remembered the tribes' decision to use some of their $18.5 million settlement (with ARCO and the state of Montana's NRDP) to restore a tributary of the Clark Fork that ran across the reservation. In particular, the tribes wanted to mend a waterway that was comparable to the headwaters of the Clark Fork.

All fifty miles of the Jocko River flow through the Flathead Indian Reservation. Resident bull trout survived in the river's upper reaches at the turn of the millennium. But its lower stretch wound through agricultural and other developed land before joining the Flathead River, which in turn flowed into the Clark Fork near the reservation's western edge. Those pressures on the lower Jocko had decimated bull trout. Using what White called a "watershed" approach, the Bull Trout Restoration Project included retooling up to twenty-two miles of the Jocko's riparian environment. Restorative efforts included improving irrigation so that more water stayed in the stream, restricting new development within the floodplain, repairing the riverbanks in places damaged by livestock grazing and development, reconstructing nearly eight hundred acres of wetlands, and replanting native plants. Those efforts mimicked the physical restoration of the Milltown area.

By the time the Clark Fork flowed past the partially dismantled Milltown Dam in 2008, the lower Jocko was meandering, pooling, and purifying in new wetlands, swirling and depositing clean sediment in handcrafted logjams and eddies, and quenching the roots of willow and cottonwood, as well as helping sustain a growing population of bull trout. As often as possible, the CSKT tried to implement what White called "passive management," a way to remove "the change agent that's causing the disturbance and allow the natural processes to restore the river. . . . The rivers always want to heal themselves." It is "the best way to restoration," according to White. In addition to pushing restoration into distant tributaries, the CSKT project emphasized education and cultural renewal.

Germaine White helped the CSKT publish *Bull Trout's Gift: A Salish Story about the Value of Reciprocity*. Based on a series of fifth-grade field trips to the Jocko, the book used the grade-schoolers' interest in the bull trout as the starting point for them to learn about the natural history of the area and the people who lived there. Illustrations accompanied a story that intertwined the modern-day Jocko with the tribes' history, traditions, and religion associated with western Montana waterways, and with changes brought about by settlement. The CSKT produced the book as part of an educational package including a field journal and an interactive DVD called *Explore the River: Bull Trout, Tribal People, and the Jocko River*. While scientific data on restoration drove much of the actual work on the Jocko, the tribes believed that lasting changes necessitated a "sense of stewardship among the community," which depended on "public information and education," in White's words.[32]

Bull Trout's Gift prompted school field trips to the Jocko, similar to the one in the book. White enjoyed seeing this offshoot of the Clark Fork Superfund site encourage lots of "experiential and observational learning" and actually "being in the water" as a way for children to study bull trout, riparian habitats, and people's relationship with the natural world. Restoring bull trout offered the gift of more and healthier native fish and a renewed connection between people and their local environments, history, and culture.

Throughout the book, readers grapple with the pronunciation and meaning of Salish names, especially those associated with rivers, wildlife, and western Montana. Those words are part of a tribal emphasis on restoring a nearly dead native language to a new generation. In 2001, the CSKT opened Nkwusm, the only Salish-language immersion school in the world. The tribe has continually expanded and revised its native-language dictionary in the last decade. Salish elder and storyteller Johnny Arlee, who has written numerous books of Salish stories and has consulted for and acted in movies such as *Jeremiah Johnson*, appears in *Bull Trout's Gift*. As the book's narrator, Arlee familiarizes his audience with important

The Clark Fork–Blackfoot confluence is known in Salish as the "place of big bull trout," named for what is now the threatened bull trout species. As part of the Clark Fork cleanup and as a way to aid replenishment of the bull trout population, the Confederated Salish and Kootenai Tribes have undertaken a large-scale watershed restoration project on the Jocko River, a tributary of the Flathead River that flows into the Clark Fork just west of the tribes' Flathead Reservation. Ronda Howlett and her fifth-grade class from Arlee gather during a field trip to the demonstration reach of the Jocko River Restoration Project. *Courtesy Confederated Salish and Kootenai Tribes, Pablo, Montana.*

Salish place names. The book reinforces this effort at restoring language with a detailed pronunciation guide. As the book's subtitle explains, it's about reciprocity. To give is to receive; to know the bull trout is to know a whole environment, a way of life, and a language.[33]

    The story also illustrated that the environment depends on people's actions. As with many Native American stories about the relationship between people and their environments, *Bull Trout's Gift* emphasizes the profound impacts humans have on the rest of the world. In this case, when a pair of Salish parents disrespected the river, it dried up, sucking the life out of the surrounding watershed. Their trek to the river's source "to make amends" for their disrespect helped coax the river back to life. *Bull Trout's Gift* was meant to acknowledge the human capacity for maintaining, destroying, and restoring environments.[34]

As part of the bull trout recovery efforts, the Confederated Salish and Kootenai Tribes created several vehicles to spotlight the importance of the bull trout to the culture of the Salish and Pend d'Oreille people: an illustrated storybook, *Bull Trout's Gift*; a field journal, *Snqeymintn*; an interactive DVD, *Explore the River: Bull Trout, Tribal People, and the Jocko River*; and a school curriculum, *Explore the River: Integrated Multimedia Curriculum Frames by the Cultural Values of the Salish and Pend d'Oreille People. Courtesy Confederated Salish and Kootenai Tribes, Pablo, Montana.*

The Clark Fork's restoration led to resuscitating other histories as well. In November 2005, Envirocon finished removing a small dam on the Blackfoot River just above the Milltown structure. As a small power-producing project and an end point for millions of board feet of logs that once flowed down the river to the Milltown and Bonner mills, it had been a precursor to the Milltown Dam. Removal of the Stimson Dam was a precursor as well, or as Envirocon's project superintendent, Brian Vibbert, said, "It's been a good practice run for Milltown Dam." That practice run evoked many retellings of the area's logging and milling history. Some of that history resurfaced physically as century-old logs, preserved by the cold water of the river, washed ashore. The restoration company plucked many of the massive logs from the river. Four of them became pillars in a new

Native American Studies building on the University of Montana campus in Missoula. Others were reserved for possible use in an interpretive center planned for the old reservoir site. Just as remediation blended into restoration, restoration was meant to promote regional economic development.[35]

~ ~ ~

By the time engineers shut off the Milltown Dam's generators for the last time on April 7, 2006, locals and the county had devised a restoration and redevelopment plan for the area. The plan included new hiking trails, a park, and an interpretive center. It also foresaw the need for setting up and funding a community council so that area residents would have an ongoing say in reshaping the confluence. The efforts of the Milltown Superfund Site Redevelopment Working Group prompted Montana senator Max Baucus to insert a $5 million redevelopment package for the group into a federal highways bill. By 2010, Baucus had garnered nearly $1 million more in federal appropriations, aside from the state NRDP's $2.6 million, to turn the confluence into a state park, which would feature a restored river environment while also attracting people to recreate.[36]

Using the river to aid the local economy seemed to make sense to everyone. Along with its dam and reservoir, Milltown was losing its namesake industry. By the summer of 2006, the Stimson Mill in Bonner had laid off most of its workforce and was nearing a complete shutdown. In a town named for producing wood products, the nationwide slump in the lumber industry meant that redevelopment was important. Rather than garnering blame for the mill's closure, the environmental effort to remove the dam became a source of economic possibility.[37]

Three decades after its passage, debates and measurements of how Superfund has impacted the national economy have focused mostly on the law's negative economic impacts—to insurance, PRPs, banks, and taxpayers, primarily. But Milltown was an example of another set of Superfund's economic consequences. Superfund projects had become part of a growing restoration economy. Just as Superfund implementation in western Montana provided a business opportunity for Envirocon, the U.S. government was embracing restoration in other ways. U.S. deputy secretary of the interior Lynn Scarlett had visited Missoula in September 2006 to deliver a keynote titled "From Resource Damages to Restoration." Speaking at the thirtieth annual Public Land Law Conference, she promoted the federal government's effort to encourage an economy of restorative work by the private sector on public land projects. By 2010, Tom Vilsack,

U.S. agriculture secretary, informed the national media that his agency's new vision "begins with restoration." The secretary singled out watersheds as particularly important to fostering the health of federal lands. The Society for Ecological Restoration, a clearinghouse for all aspects of the restoration economy, from jobs to conferences to major corporations, began in 1988 and counted members in more than seventy countries by 2010. Using national parks as case studies, Washington University in St. Louis professor William Lowry argued in his 2009 book *Repairing Paradise* that by the twenty-first century Americans and their government had realized that preserving renowned environments was not enough. Even many of the country's most cherished landscapes warranted restoration. By 2012, American Rivers, which had once ranked the Blackfoot River at the top of its most endangered waterways list, launched a campaign to remove one hundred dams and restore the sections of rivers they impounded within a year. The group listed Milltown as a model.[38]

Besides serving as an emblem in a national trajectory, fixing environmental damage at Milltown held potential for local entrepreneurs as well. In the heat of the summer of 2005, a new tube rental and river shuttle service was thriving along the Blackfoot. Fly-fishing guides reported that the improved image of the rivers associated with restoration (even if most of it was yet to happen) was boosting business. State economic reports showed that 39 percent of all guided trips in Montana were on rivers, amounting to more than $50 million a year. A 1999 study showed that tourism brought $150 million to Missoula County each year. By the spring of 2006, the *Missoulian* reported that a two-day event called the "Governor's Restoration Forum," spurred by the magnitude of work on the Clark Fork, was "the first comprehensive look at the economic and public benefits of revitalizing Montana's landscapes." Storm Cunningham, the keynote speaker, had recently authored *The Restoration Economy: The Greatest New Growth Frontier*. Positioning themselves on that frontier, a Milltown couple reopened a defunct whitewater rafting company in 2007. In the small riverside town of Turah, just upstream from the doomed dam, Kathy Marshall, owner of the hamlet's single small store, credited the Superfund cleanup at Milltown with renewing her hopes for staying in business.[39]

As UM economist Thomas Power argued about the value of place, "attractive natural landscapes influence the location of economic activity." While Power studied preserved places, such as national parks and federal wilderness areas, to make his point, what was happening on the Clark

Fork demonstrated that reviving damaged environments led to similar economic benefits.[40]

~ ~ ~

There were many immediate and tangible qualities to remediation, restoration, and redevelopment at Milltown. Millions of tons of toxic sediment were gone. The river flowed in a meandering channel feeding native riparian plants and depositing sediment that fostered aquatic life rather than killing it. Businesses sprang up in hopes of capturing clientele attracted to fishing and floating opportunities on the renewed rivers. Groundwater models predicted that the area's aquifer would rid itself of arsenic in four to ten years.[41]

Yet in some ways, the removal of Milltown Dam and the area's restoration was an experiment, the outcome unknown. The renewal of language study on the Flathead Reservation was one such trial. Another was a series of art exhibits at four Missoula museums in the autumn of 2006. The art ranged from photographs of the rebounding rivers to a branch and cloth sculpture that looked a bit like clean water flowing over skeletal hands, perhaps signifying new life emerging from a dead past. The most notable of the exhibits in the "Changing Currents" show was a room-sized installation piece. German-born "internationally renowned artist" Gerhard Trimpin constructed a wall map of the Clark Fork and marked it with significant river restoration events and places. He wired the map so that when a sensor mounted on rails glided over each mark, it closed an electric circuit that activated the lifting and lowering of bamboo tubes within vats of water placed throughout a room of the Missoula Art Museum. Cut to precise lengths, the tubes hummed different notes as their movement blew air over reeds within the segments of bamboo. Trimpin's installation, like all of the art, was an abstract way of pondering the future of a restored river. As critic Betsy Cohen wrote, the art asked such questions as, "What will the river sound like? What new paths will it forge? How wild will it be?"[42]

Answers to those questions came in small steps. In late 2006, tiny trout swam in Butte's Silver Bow Creek for the first time in nearly a century. The *Missoulian* likened the fish discovered by a Montana FWP electroshocking project to a "canary in a coal mine in reverse," the fish an indicator of a rejuvenating stream. It took three more years before someone actually caught a trout from that stream using rod and reel and another three before the state published a fishing report for the Clark Fork's headwaters. At Milltown, the river was flowing through its newly created "historic" channel by then. In 2010 the state FWP was buying land around the confluence of the Blackfoot and Clark Fork, as well as on the overlooking bluff, the site of so many celebrations. After giving a parcel to the Bonner school, making it "possibly . . . the first elementary school in the nation with its own community school forest," according to superintendent Doug Ardiana, the state aimed to gather community input to steer its plans for making the land into a state park. Superfund money from ARCO funded the purchase and development. With the dam's removal playing no small part, fisheries biologists declared the bull trout recovered in places like the Clearwater River, one of the Blackfoot's last major tributaries before it joins the Clark Fork.[43]

But the Superfund pocket filled by ARCO was not bottomless. Within two years of the dam's removal, the number and diversity of proposals for restoration work strained the fund and its management. As Peter Nielsen said of projects that ranged from museums to trail work far from the Clark Fork, "there's not enough money." Although fixing a problem with Milltown's sediment in Opportunity Ponds demanded more money from ARCO, the proposed work received swift approval from the EPA, river advocates, and the community. As of February 2012, the Milltown soil had failed to grow much vegetation on nearly seven hundred acres of ARCO's upper Clark Fork repository, so the company had to continually mitigate dust blowing off the site. Having pushed the decision to put the sediment in Opportunity, all the involved parties agreed on as quick, thorough, and permanent a solution to the bare site as possible. ARCO consented to pay another $15 million to cap the Milltown deposit with a combination of lime, fresh topsoil, and a new round of plantings.[44]

Superfund implementation at Milltown demonstrated the efficacy of combining remediation, restoration, and redevelopment, as well as making community involvement an integral part of that combination. Having helped write the EPA's first guidelines for developing community advisory groups at Superfund sites, Diana Hammer saw her work as the community coordinator at Milltown helping to push the agency in that direction. She saw the measure of success at the confluence of the Blackfoot and Clark Fork as "the opportunity to take a federal Superfund site and turn it into a state park." That transformation echoed the progress in the law from its inception as an emergency waste cleanup measure to one that, at best, made cleaning up toxic waste one step in renewing damaged landscapes and revitalizing economies.[45]

The original Milltown sediments failed to grow as much cover crop as was expected, so ARCO funded laboratory tests to determine how to amend the sediments. The addition of lime and extra soil was the roughly $15 million solution that led to the results shown in this 2014 view across part of the Opportunity Ponds reclamation area. Native grasses now grow in the reclaimed Opportunity Ponds that originally held tailings from the Anaconda Company's former smelter operations in nearby Anaconda. Such wildlife as elk, moose, deer, antelope, and birds now take advantage of the area, including the sandhill cranes and antelope shown here. The tower in the background is left over from the years that the company dumped tailings, in a slurry, from the smelter into the ponds. Called a decant tower, the structure aided in removing water from the slurry to make room for more tailings. *Courtesy Kenneth Brockman, photographer, Bureau of Reclamation, Butte, Montana.*

A year before restoration of Milltown was completed, the EPA published a case study of the site "to provide relevant information and lessons learned for parties interested in Superfund site reuse, river restoration and recreational and ecological land revitalization." The emphasis of that case study was the ways in which local residents and community groups had shaped remediation, promoted restoration, and, most uniquely, linked them to economic and recreational redevelopment around the site. Because of Milltown, the linkage of remediation, restoration, and redevelopment became the "3 Rs" in EPA jargon and its watermark for Superfund cleanup. The top two lessons learned in the EPA case study of Milltown were about generating and responding to community engagement. As for the innovation of linking restoration and redevelopment to remediation, this case study included the reflections of a former EPA Superfund site manager that "at older sites, EPA did not focus on taking reuse considerations into account early in the cleanup process." The "3 Rs" at Milltown were part of an evolution in Superfund implementation from its origins as a toxic waste containment program to one with the goal of thorough site restoration and economic rehabilitation. Or, as the case study phrased it, "thanks to lessons learned at Superfund sites like the Milltown Reservoir Sediments site, EPA has developed additional tools to ensure an integrated approach to the cleanup and redevelopment of contaminated lands."[46]

Montana skiers loved the winter of 2010–2011. Slopes reported near-record snowpack and lift ticket sales. Come spring, river watchers were more anxious than excited about all that snow. With high flood levels a distinct possibility, Envirocon workers bulked up protection of their restoration work. But a lot would depend on the river. Running at the "35-year flood" level, the Clark Fork flushed nearly fourteen thousand cubic feet per second through Milltown after gaining the Blackfoot's flow. When runoff finally settled in midsummer, the results were encouraging. Although the flood waters tweaked some of the newly built S curves upriver from the old dam site, the restoration work held, signifying that what Envirocon had constructed was acting like a fairly natural floodplain. Once again, people celebrated.[47]

Conclusion

Men may dam it and say that they have made a lake, but it will still be a river. It will keep its nature and bide its time, like a caged animal alert for the slightest opening. In time, it will have its way; the dam, like the ancient cliffs, will be carried away piecemeal in the currents.

—Wendell Berry

In a mucked up lovely river,
I cast my little fly.
I look at that river and smell it
And it makes me want to cry.
Oh to clean our dirty planet,
Now there's a noble wish,
And I'm puttin' my shoulder to the wheel
'Cause I wanna catch some fish.

—Greg Brown, "Spring Wind," *Dream Café*

What happened at Milltown didn't tip the balance of any presidential elections, nor did it birth or overturn any major laws. Few national environmental organizations got involved. On the rare occasions they did, it was to lend token support to what was already transpiring. What happened at Milltown was that people participated in the implementation of a major environmental policy over the course of more than thirty years. They engaged in the kind of process that was more common than extraordinary. Rather than laying claim to shaping a rare sea change, they made the kind of history that happens most every day by affecting incremental change from within the bounds of the law. It is tempting to compare the patient, often tedious, work people undertook and the unexpected changes that work bore to the slow, steady forces of a river that, as Wendell Berry noted, "will have its way."

One of the things that made the Superfund process at Milltown important was its commonness. The EPA has by now considered tens of thousands of places as possible Superfund sites. At any one time since the law's inception, a few hundred to a few thousand of these make it onto the agency's annual National Priorities List. The majority of them share qualities with the Milltown site, such as their location on the fringes of urban centers, industrial history (especially mining), contribution to water (especially groundwater) contamination, and the presence of arsenic or other heavy metals as their primary human health risk. They range in size from about the area of an average suburban backyard to the 120 miles of river and riparian area that made the Clark Fork Superfund Complex the largest such site in the country. Every state has them, and even in a large, thinly populated place like Montana, Superfund sites are usually within an hour or so drive of any point on a map. In other words, just about everyone in the United States lives near their own Milltown. And the cleanup that happened at Milltown, as well as its impacts on Superfund implementation, was in large part thanks to local residents.[1]

Milltown had the advantage of being just upstream from a community with a long history of social activism, of which environmental advocacy was a developing strong suit. Missoulians spearheaded public participation at Milltown, where more people wrote to the EPA about the site than ever before and often did so outside the designated comment period. That response to Milltown encouraged the EPA to embrace more and earlier public input. The EPA now maintains a community involvement link on its Superfund website. The agency helps fund and organize half a dozen kinds of community groups meant to help citizens engage in the Superfund process. And each year since 2000, EPA administrators have selected an individual or group to receive the "Citizen's Excellence in Community Involvement Award." People's participation at Milltown did not single-handedly bring about all these changes. But as Milltown's EPA community coordinator, Diana Hammer, recognized, "cleanup [at Milltown] was

Many quantifiable outcomes have resulted from the Milltown remediation and restoration. Millions of tons of toxic sediment have been removed, the area's aquifer should be rid of arsenic in a few years, the obstacle to fish migration is gone, new meanders and wildlife habitat have been restored to the Clark Fork, and a new state park has been created to provide public access and space for recreation. This aerial view (looking southeast) shows the relationship of the town, the newly winding river, the confluence, Interstate 90, the railroad tracks, and the local roads. Milltown State Park is located beside the confluence of the two rivers—in the wedge between the interstate and railroad tracks in this view—and wraps under the bridges of the interstate, train, and frontage road to follow the Blackfoot upriver opposite Milltown. There's an auxiliary site on the bluff across the Clark Fork, and a proposed pedestrian bridge will cross the river. *Courtesy Judy Matson, photographer, and Gary Matson, pilot.*

unique and a great Superfund model because of all the involvement of informed advocates. . . . The EPA does a better job of this now than we used to [of] being more upfront with people about what's going on and letting them look behind the curtain and seeing how are those decisions being made and actually be part of it."[2]

Because public input promoted, if not demanded, restoration, Milltown's outcome was part of a shift in Superfund toward a greater concern for restoring natural environments, instead of simply removing or containing toxins. At Milltown that meant getting the worst toxic sediments out of the floodplain. Equally important, the EPA dovetailed sediment removal, a traditional remediation, with the creation of a naturally functioning river channel that is beginning to improve water quality and provide a variety of fish and wildlife habitat, from riparian trees and shrubs to wetlands and modest river rapids. Superfund money generated through the law's corporate liability provisions has helped pay for some of the law's unexpected consequences. Restoration of bull trout habitat, as well as native language recovery efforts on the Flathead Indian Reservation, made that point. Those unexpected types of restoration also revealed how the law's implementation encouraged tribal participation in issues off the reservation that were of broad public interest.

In addition, the EPA used remediation and restoration as springboards to economic redevelopment. At the turn of the millennium, cleaning up and restoring the Milltown site became one of Envirocon's largest and most lucrative undertakings. For the first time along the Clark Fork, enhancing rather than extracting from the natural environment became a viable economic driver. The fix at Milltown was meant to create long-term financial opportunities as well. Conversion of nearly five hundred acres of land once owned by the local mills, much of which had spent one hundred years underwater, into Milltown State Park is one such example. A place once deemed a toxic waste site of national concern will become a healthy environment for people to congregate, recreate, and, ostensibly, spend money in the local community. The new state park, like so much at Milltown, resulted from public input and involvement.

Although Superfund implementation at Milltown was shaped in large part by the diligent participation of local residents, as with so much of history, contingencies helped along the way. As Wendell Berry discerned, the Clark Fork kept "its nature" and bided "its time." A few spring floods and ice jams heightened attention to the dam's age and safety and attuned people's attention to the river's ailing health beyond the defined boundaries of the original Superfund site on the Clark Fork. Bull trout and other native fishes continued to follow their evolutionary impulses upstream in search of ancient spawning waters even though a human-made collection of rock, timber, concrete, and steel had blocked their way for a century.

The presence of fish thwarted in their spawning efforts provided biologists and concerned citizens with added justification for removing Milltown Dam. When a nonnative northern pike in Milltown Reservoir ate Pollywog, a cutthroat trout radio-tagged by a fisheries biologist and tracked by a group of grade school students in the area, it educated a growing contingent of people about yet another way the Superfund site was wreaking havoc on native fish.

Other contingencies were of a more human nature. Peter Nielsen's inadvertent discovery of classified documents about Milltown on a FERC web page, which someone in an office thousands of miles away posted mistakenly, harnessed post-9/11 fears of terrorism to the more tangible concerns about the Milltown Dam's stability and downstream impacts. When an acquaintance bumped into Tracy Stone-Manning in the grocery store and suggested a Milltown to Downtown float, he added a powerful demonstration of community and recreational concern to her organization's dam removal and river restoration campaign. Admitting that it might be a "corny" expression, Diana Hammer said that her colleagues "joked that it's not only a project at the confluence of two rivers but it's been a confluence of people and a confluence of events," some purposeful, some not.[3]

Most of what happened at Milltown resulted from how a community involved itself in the Superfund process. Local scientists connected arsenic pollution in drinking water to the reservoir and the long history of upstream mining. People who moved to Missoula to attend the university stayed and committed their lives to working for local government and environmental organizations that shaped the outcomes at Milltown. Local business owners devoted money and time to the same causes. A record number of citizens wrote to the EPA supporting the eventual solution of removal and restoration. Opponents of the plan pushed the cause of economic redevelopment. As Greg Brown wrote in his song "Spring Wind," people participated in the Superfund process because of very personal cares and ways of knowing the Clark Fork watershed.

Then there are outcomes that may still be in the making and will not be found in any law, state park, or river. When asked if Milltown is important beyond the local communities and local environmental organizations, Trout Unlimited state director Bruce Farling, who was active in the site's implementation from nearly the beginning, said, "This is a pretty big deal. . . . I think this tells other communities what the possibilities are. They don't have to settle for lowball stuff. . . . I mean a $140 million cleanup and restoration . . . against two very powerful corporations." Farling's sentiments fit with those of Jenny Price's 2008 call for river restoration to become the icon for twenty-first-century environmentalism. Both of them seem to understand that just as many places, particularly watersheds, need serious revitalization, people need stories of an improving environment as much as they need reminders of its degradation. But what happened at Milltown is more than a single story or icon. Practical restoration of damaged environments represents the heart of American environmentalism at the turn of the millennium.[4]

Three years after the combined flow of the Clark Fork and Blackfoot River scoured its way through a breach in Milltown Dam, the river's course through Hellgate Canyon ran nearly unimpeded. Footings for Interstate 90 still stood in the Blackfoot, just above the confluence. I-90 still followed the river's path through most of western Montana. Old logs from Milltown's former life as the major industrial processing center between forest and finished timber jutted onto the nearby banks. The stumps of trees cut more than a century ago studded the old reservoir bed. Restoration did not mean that the Clark Fork would look like it did when the Salish gave it names like "place of the big bull trout," or "shining waters." It did mean that arsenic levels were dropping in the groundwater. Native plants were sprouting on the riverbanks and nonnative fish populations were dwindling. So, too, native fish swam freely past the old dam site. And although it was closed to recreational traffic for at least another year, a few rogue boaters floated through the confluence on nights when moonlight allowed them to cast flies for trout, see the shine of water tumbling over rocks, and enjoy a stretch of river no one had plied for generations. After spring high water subsided in 2014, the Blackfoot and Clark Fork confluence was fully open to all floaters for the first time in more than a century.[5]

Notes

Introduction

1. Elizabeth Grossman, *Watershed: The Undamming of America* (New York: Counterpoint, 2002), 3.

2. "Superfund: Cleaning Up the Nation's Hazardous Wastes Sites," U.S. Environmental Protection Agency, accessed 9/9/2014, http://www.epa.gov/superfund/index .htm; Katherine D. Walker, March Sadowitz, and John D. Graham, "Confronting Superfund Mythology: The Case of Risk Assessment and Management," in *Analyzing Superfund: Economics, Science, and Law*, ed. Richard L. Revesz and Richard B. Steward (Washington, D.C.: Resources for the Future, 1995), 40–41; Katherine N. Probst, "Evaluating the Impact of Alternative Superfund Financing Schemes," in *Analyzing Superfund*, 148–49; Jack Lewis, "Superfund, RCRA, and UST: The Clean-Up Threesome," *EPA Journal* 17, no. 3 (July/August 1991): 10; Katherine N. Probst, Don Fullerton, Robert E. Litan, and Paul R. Portney, *Footing the Bill for Superfund Cleanups* (Washington, D.C.: Brookings Institution and Resources for the Future, 1995); William R. Lowry, *Dam Politics: Restoring America's Rivers* (Washington, D.C.: Georgetown University Press, 2003), 32–33.

3. Norman Maclean, *A River Runs through It* (New York: Simon and Schuster, 1976), 13; William Farr, "Going to Buffalo: Indian Hunting Migrations across the Rocky Mountains. Part 1, Making Meat and Taking Robes," *Montana: The Magazine of Western History* 53, no. 4 (Winter 2003): 2–21; William Farr, "Going to Buffalo: Indian Hunting Migrations across the Rocky Mountains. Part 2, Civilian Permits, Army Escorts," *Montana: The Magazine of Western History* 54, no. 1 (Spring 2004): 26–43; Richard Manning, *One Round River: The Curse of Gold and the Fight for the Big Blackfoot* (New York: Henry Holt, 1997).

4. Edwin Dobb, "Pennies from Hell," *Harper's Magazine*, October 1996.

5. Michael P. Malone, Richard B. Roeder, and William L. Lang, *Montana: A History of Two Centuries*, rev. ed. (Seattle: University of Washington Press, 1976); Michael

Malone, *The Battle for Butte: Mining and Politics on the Northern Frontier, 1864–1906* (Helena: Montana Historical Society Press, 1981); David Stiller, *Wounding the West: Montana, Mining, and the Environment* (Lincoln: University of Nebraska Press, 2000). For specifics on Milltown's development, see also Shirley Coon, "The Economic Development of Missoula, Montana," (master's thesis, University of Chicago, 1926). For area geology, see David D. Alt and Donald W. Hyndman, *Roadside Geology of Montana* (Missoula, Mont.: Mountain Press, 1986).

6. Malone, Roeder, and Lang, *Montana*.

7. George F. Weisel, "Ten Animal Myths of the Flathead Indians," *Anthropology and Sociology Papers* 18 (1959): 9; Ella E. Clark, *Indian Legends from the Northern Rockies* (Norman: University of Oklahoma Press, 1966): 61–136; Salish–Pend d'Oreille Culture Committee and Elders Cultural Advisory Council, Confederated Salish and Kootenai Tribes, *The Salish People and the Lewis and Clark Expedition* (Lincoln: University of Nebraska Press, 2005). For Missoula history, see Coon, "Economic Development of Missoula."

8. For the development of Missoula, see Coon, "Economic Development of Missoula"; Ruth Boydston Scott, *Missoula: Trading Post to Metropolis* (Missoula, Mont.: R. Scott, 1997); and Malone, Roeder, and Lang, *Montana*.

9. Liza Nicholas, Elaine M. Bapis, and Thomas J. Harvey, eds., *Imagining the Big Open: Nature, Identity, and Play in the New West* (Salt Lake City: University of Utah Press, 2003); David Brooks, *Bobos in Paradise: The New Upper Class and How They Got There* (New York: Simon and Schuster, 2000), 104.

10. "Missoula Nonprofits," Make It Missoula, http://www.makeitmissoula.com /community/non-profits/; Jo Rainbolt, *Missoula Valley History* (Dallas: Curtis Media, 1991); *Montana Constitutional Convention, 1971–1972* (Helena: Montana Legislature, 1979–1982); Bicentennial Committee Bonner School, *A Grassroots Tribute: The Story of Bonner, Montana* (Missoula, Mont.: Gateway Press, 1976).

11. For open space information, see "Missoula's Open Space Program: Making Missoula a Better Place," accessed 4/2/2012, http://www.ci.missoula.mt.us/Document View.aspx?DID=650; "Missoula Montana: Open Space," City of Missoula, accessed 4/2/2012, http://www.ci.missoula.mt.us/index.aspx?NID=185.

12. Jenny Price, "Remaking American Environmentalism: On the Banks of the L.A. River," *Environmental History* 13 (July 2008): 536–55 (quote on 539); Steven Hawley, *Recovering a Lost River: Removing Dams, Rewilding Salmon, Revitalizing Communities* (Boston: Beacon Press, 2012); Dan Flores, *The Natural West: Environmental History in the Great Plains and Rocky Mountains* (Norman: University of Oklahoma Press, 2001). Also see Dave Forman, *Rewilding North America: A Vision for Conservation in the 21ˢᵗ Century* (Washington, D.C.: Island Press, 2004); and Christopher McGrory Klyza and Bill McKibbin, eds., *Wilderness Comes Home: Rewilding the Northeast* (Middlebury, Vt.: University Press of New England, 2001).

13. Lowry, *Dam Politics*, 2–3, 17, 59, 83–87, 225.

14. Shannon Petersen, *Acting for Endangered Species: The Statutory Ark* (Lawrence: University Press of Kansas, 2002); Matthew J. Lindstrom and Zachary A. Smith, *The*

National Environmental Policy Act: Judicial Misconstruction, Legislative Indifference, and Executive Neglect (College Station: Texas A&M University Press, 2001); Revesz and Steward, *Analyzing Superfund*; U.S. EPA, Region 8, "Citizens' Superfund Workshop," 5/5/1990, reel 241, Milltown Administrative State Site Records, U.S. Environmental Protection Agency, Government Documents, Mansfield Library, University of Montana (subsequent citations from this source will appear as "Milltown Administrative State Site Records").

15. Bruce Farling interview, OH 428–04, David Brooks's Milltown Oral History Project, 2009–2011, Archives and Special Collections, Maureen and Mike Mansfield Library, University of Montana, Missoula (subsequent citations from this source will provide the OH number only).

Chapter I

1. William Woessner OH 428–01, 9/22/2009; William W. Woessner and Marin A. Popoff, "Hydrogeologic Survey of Milltown, Montana and Vicinity," University of Montana, Department of Geology, Missoula, 6/15/1982; letter from Johnnie Moore and William Woessner to Missoula City-County Health Department, 10/3/1982, in Milltown Collection at MCCHD. Also see Sherry Devlin, "History's Troubles," *Missoulian* (Missoula, Mont.), 1/27/2002.

2. Leo and Theola Dufresne, OH 419–06, Milltown Oral History Collection, Archives and Special Collections, Mansfield Library, University of Montana, Missoula, (subsequent citations from this source will provide the OH number only); Dick Manning, "Safe Water Finally Flows in Milltown," *Missoulian*, 5/10/1985; Kevin Miller, "Arsenic Probe Is Stepped Up; Warning Issued," *Missoulian*, 12/16/1981; David Roach, "Contamination Leaves Residents Perplexed," *Missoulian*, 12/16/1981. Also see Johnnie N. Moore and William W. Woessner, "Arsenic Contamination in the Water Supply of Milltown, Montana," in *Arsenic in Ground Water: Geochemistry and Occurrence*, ed. Alan H. Welch and Kenneth G. Stollenwerk (Norwell, Mass.: Kluwer Academic Publishers, 2003), 331.

3. Peter Neilsen, Missoula City-County Health Department, personal communication, 10/14/2009; notes in author's possession.

4. Kevin Miller, "Arsenic Found in Milltown Water Supplies," *Missoulian*, 12/15/1981; Kevin Miller, "Arsenic Probe Is Stepped Up; Warning Issued," *Missoulian*, 12/16/1981; David Roach, "Contamination Leaves Residents Perplexed," *Missoulian*, 12/16/1981.

5. One gram equals 1,000,000 µg, and roughly 1,000 grains of sugar. Therefore, 370 µg of sugar equal 0.37 grains. One gallon equals 3.78 liters. So, 370 µg/L, or 0.37grains/L, of sugar equal 1.4 grains/gallon.

6. Kevin Miller, "Arsenic Probe Is Stepped Up; Warning Issued," 12/16/1981; David Roach, "Contamination Leaves Residents Perplexed," *Missoulian*, 12/16/1981; David Roach, "Officials Test More Wells," *Missoulian*, 12/17/1981; editorial, "Arsenic Episode Revealed Failings," *Missoulian*, 12/13/1981; Kevin Miller, "Quantity and Toxicity of Arsenic in Milltown Top Previous Levels," *Missoulian*, 12/22/1981;

Kevin Miller, "Hair, Fingernails Tests Planned in Arsenic-Contamination Probe," *Missoulian*, 12/23/1981; Jose M. Azcue and Jerome O. Nriagu, "Arsenic: Historical Perspectives," in *Arsenic in the Environment, Part 1: Cycling and Characterization*, ed. Jerome O. Nriagu (New York: John Wiley and Sons, 1994).

7. David Roach, "Contamination Leaves Residents Perplexed," *Missoulian*, 12/16/1981.

8. U.S. EPA, "This Is Superfund: A Citizen's Guide to EPA's Superfund Program," March 1994, 1, National Service Center for Environmental Publications, http://nepis .epa.gov.

9. Carole Stern Switzer and Peter Gray, *CERCLA: Comprehensive Environmental Response, Compensation, and Liability Act (Superfund)*, 2nd ed. (Chicago: American Bar Association, Section of Environment, Energy, and Resources, Basic Practices Series, 2008), 1–14; William K. Reilly, "A Management Review of the Superfund Program," 1989, National Service Center for Environmental Publications, http://nepis .epa.gov.

10. Thomas R. Wellock, *Preserving the Nation: The Conservation and Environmental Movements, 1870–2000* (Wheeling, Ill.: Harlan Davidson, 2007); Samuel P. Hays, *Beauty, Health, and Permanence: Environmental Politics in the United States, 1955–1985* (Cambridge: Cambridge University Press, 1987); Lois Marie Gibbs, *Love Canal and the Birth of the Environmental Health Movement* (Washington, D.C.: Island Press, 2011).

11. Coon, "Economic Development of Missoula." Note that the official name of the Anaconda Company changed over time. It has been variously known as the Anaconda Copper Mining Company, the Anaconda Copper Mining and Smelting Company, the Anaconda Mining Company, or simply "The Company." For simplicity, this volume will use "Anaconda Company."

12. Ibid., tables 18 and 27.

13. Dufresne, OH 419–06; Mildred Miller Dufresne, OH 140–29, Montanans at Work Oral History Project, Archives and Special Collections, Mansfield Library, University of Montana, Missoula, (subsequent citations from this source will provide the OH number only); *Church Decorated for Christmas, Norwegian Heritage Picnic*, photoboards 127 and 128, Jack L. Demmons / Bonner School Photographs, Archives and Special Collections, Mansfield Library, University of Montana, Missoula, http://www .lib.umt.edu/digital/demmons; Bicentennial Committee Bonner School, *Grassroots Tribute*. For general interpretation on immigration see John Bodnar, *The Transplanted: A History of Immigrants in Urban America* (Bloomington: Indiana University Press, 1985). See also American Social History Project, *Who Built America? Working People and the Nation's Economy, Politics, Culture, and Society*. Vol. 2, *From the Gilded Age to the Present* (New York: Pantheon Books, 1992).

14. Dufresne, OH 419–06. For similar stories of growing up in Milltown see the full collection of oral histories by Minnie Smith and Caitlin DeSilvey, OH 419–01–08. Also see Patrick Thibodeau, OH 419–23/24; and *Blackfoot River near the Chaffey*

Homestead, photoboard 44, 1937; *Skating on the Frozen Reservoir below Milltown,* photoboard 51, undated; *Sailboat on Milltown Dam Reservoir,* photoboard 116, undated; *Clark Fork River Swimming Hole,* photoboard 166, undated; *Hazel Beadle at the Swimming Hole,* photoboard 166, undated; *On the Blackfoot River,* photoboard 166, undated; and *Orvo and Ellen Elo on the Bank of the Blackfoot,* photoboard 199, 1939; in Jack L. Demmons / Bonner School Photographs.

15. Dufresne, OH 419–06; John Price, OH 419–07. For other documents on fishing, see *Four Fishermen Display Their Catch,* photoboard 75; and *Ice Fishing on the Blackfoot River,* photoboard 195, in Jack L. Demmons / Bonner School Photographs.

16. Thibodeau, OH 140–23/24; Dufresne, OH 140–29; Arthur Lehti, OH 140–30; Bicentennial Committee Bonner School, *Grassroots Tribute,* 52.

17. Kevin Miller, "Arsenic Found in Milltown Water Supplies," *Missoulian,* 12/15/1981.

18. Dennis Pleasant, OH 419–04; Woessner, OH 428–01; William W. Woessner and Marin A. Popoff, "Hydrogeologic Survey of Milltown, Montana and Vicinity," Department of Geology, University of Montana, Missoula, 6/15/1982; Johnnie Moore and William Woessner to Missoula City-County Health Department, 10/3/1982 (original copy in Milltown Collection, MCCHD). Also see Sherry Devlin, "History's Troubles," *Missoulian,* 1/27/2002; William W. Woessner, Johnnie N. Moore, Carolyn Johns, Marin A. Popoff, Leslie C. Sartor, and Mary Lou Sullivan, *Final Report: Arsenic Source and Water Supply Remedial Action Study, Milltown, Montana* (Department of Geology, University of Montana, Missoula, 7/31/1984); and Ann K. Bailey, "Concentration of Heavy Metals in the Sediments of a Hydroelectric Impoundment" (master's thesis, University of Montana, 1976). Another survey, conducted by Associate Professor Jerry Bromenshenk in the Environmental Studies Department at the University of Montana, found arsenic levels far exceeding those in Milltown in soil and vegetation samples taken from "denuded areas along the Clark Fork River" above the reservoir. Bromenshenk believed that that arsenic, like Milltown's contamination, was from upriver mining and smelting. Letter from Jerry Bromenshenk to Ms. Elaine Bild, director of environmental health, Missoula City-County Health Department, 11/17/1982, in Milltown Collection at MCCHD.

19. Woessner, OH 428–01; letter from Johnnie Moore and William Woessner to Missoula City-County Health Department, 10/3/1982 (original copy in Milltown Collection, MCCHD). Also see Sherry Devlin, "History's Troubles," *Missoulian,* 1/27/2002; and Woessner et al., *Final Report.*

20. Moore and Woessner, "Arsenic Contamination," 329–50 (quote from 334).

21. Woessner, OH 428–01; Woessner and Popoff, "Hydrogeologic Survey of Milltown"; Woessner et al., *Final Report.*

22. "Superfund: Basic Information," U.S. Environmental Protection Agency, http://www.epa.gov/superfund/about.htm.

23. Studies of public perceptions in the early 1990s showed that the overwhelming majority of people saw "water pollution and toxic waste sites" (77 and 72 percent,

respectively) as the most serious threats to the environment. Robert W. Adler, Jessica C. Landman, and Diane M. Cameron, *The Clean Water Act 20 Years Later* (Washington, D.C.: Island Press, 1993), 2, 11. Also see Hays, *Beauty, Health, and Permanence*; and Hal K. Rothman, *The Greening of a Nation: Environmentalism in the United States since 1945* (Orlando: Harcourt, Brace, 1998).

24. Woessner, OH 428–01; Hays, *Beauty, Health, and Permanence*, 198–206.

25. U.S. EPA, "Documentation Records for Hazardous Ranking System, June 28, 1982," reel 144, Milltown Administrative State Site Records; Moore and Woessner, "Arsenic Contamination," 329–50.

26. David Roach, "Land Given to Milltown Association," *Missoulian*, 5/20/1983.

27. Letter from Edward Zuleger, Missoula City-County Health Department, to Elaine Bild, in Milltown Collection at MCCHD; "Minutes of Milltown Water Users Association, 5/12/1982," reel 144, Milltown Administrative State Site Records; Edwin Bender, "Contamination," *Missoulian*, 5/8/1982, Edwin Bender, "Group Gears Up to Solve Water Problem," *Missoulian*, 5/14/1982; Edwin Bender, "Champion Helps Milltown Group Trying to Solve Water Problem," *Missoulian*, 5/28/1982; Steve Woodruff, "Arsenic Found in Sediment at Milltown Dam," *Missoulian*, 10/3/1982; no by-line, "Candidate Wants Arco to Fund Study of Arsenic," *Missoulian*, 10/5/1982; David Roach, "Land Given to Milltown Association," *Missoulian*, 5/20/1983; "Action Line," *Montana People's Action Newsletter* 1, no. 1 (August 1983).

28. U.S. EPA, Region 8, "Citizens' Superfund Workshop," 5/5/1990, reel 241, Milltown Administrative State Site Records; Walker, Sadowitz, and Graham, "Confronting Superfund Mythology."

29. Steve Woodruff, "Milltown Contamination Moved Up on EPA List," *Missoulian*, 10/26/1982; editorial, "Ticklers: Hope Improves for Milltown," *Missoulian*, 11/7/1982.

30. Steve Woodruff, "Milltown Arsenic on EPA List for Cleanup," *Missoulian*, 12/21/1982; editorial, "No Substitute for a Stitch in Time," *Missoulian*, 12/27/1982; reader comment, "Help Milltown with Pollution," *Missoulian*, 2/11/1983; letter from Ted Schwinden to Missoula Board of County Commissioners, EPA, Department of Health and Environmental Sciences, and Department of Natural Resources and Conservation (copy in author's possession); letter from Max Baucus to Steve Durham, 11/17/1982 (copy in author's possession); letter from Missoula Board of County Commissioners to Russel H. Wyer, 1/12/1983 (copy in author's possession). For information on reservoir drawdowns, see Peter Neilsen, Missoula City-County Health Department, personal communication, 10/14/2009 (notes in author's possession); letters opposed to Montana Power Company's drawdowns, reel 465, Milltown Administrative State Site Records; also, Sherry Devlin, "Plan to Flush Milltown Lake Gets Cool Reception," *Missoulian*, 1/24/1981; Mike McInally, "Lawsuit over Milltown Dam Drawdown is Dismissed," *Missoulian*, 1/19/1982.

31. Lisa Fleischer, *Health Study of Milltown, Montana* (Missoula, Mont.: Mont-PIRG, 1983); letters to the EPA, reel 144, Milltown Administrative State Site Records.

32. Steve Woodruff, "Milltown Contamination More Urgent, Official Says," *Missoulian*, 3/9/1983; no byline, "Bill Signed to Provide Cleanup Funds," *Missoulian*, 3/29/1983; House Bill 200, 48th Legislature, 1983 (copy in possession of author).

33. Steve Woodruff, "Milltown Cleanup Pact Near," *Missoulian*, 5/3/1983; Steve Woodruff, "EPA Frees $570,000 for Milltown Aid," *Missoulian*, 7/2/1983. Also see "NPL Site Narrative for Milltown Reservoir Sediments," http://www.epa.gov/super fund/sites/nplsnl/n0800445.pdf.

34. Michael S. Gorby, "Arsenic in Human Medicine," in *Arsenic in the Environment, Part 2: Human Health and Ecosystem Effects*, ed. Jerome O. Nriagu (New York: John Wiley and Sons, 1994).

35. For the EPA's efforts regarding toxic contamination of water through the 1970s, see Hays, *Beauty, Health, and Permanence*, 198–206.

Chapter 2

1. "Missoula," Climate-zone.com, accessed 3/10/2013, http://www.climate-zone. com/climate/united-states/montana/missoula/; "Fast Climate Facts: Missoula, Montana," U.S. Department of Commerce, National Oceanic and Atmospheric Administration, National Weather Service, accessed 1/18/2011, http://www.wrh.noaa.gov/mso /climfacts.php; "USGS Water Resources: USGS Surface-Water Monthly Statistics for Montana," U.S. Geological Survey, accessed 10/2/2014, http://waterdata.usgs.gov/mt /nwis/monthly/.

2. John Stromnes and Larry Howell, "High Water: Warm Weather Poses Problems for Missoulians," *Missoulian*, 2/25/1986.

3. Michael Moore, "Missoula Mops Up," *Missoulian*, 2/26/1986.

4. Steve Woodruff, "Dam, Suckers Survive," *Missoulian*, 4/5/1986.

5. Peter Nielsen, "Milltown Dam History: Chronology Compiled by Missoula City-County Health Department," 12/3/2001, digital copy in author's possession.

6. Letters from Jerry Rourke to A. H. Wethey, 10/10/1905, 10/12/1905, 2/5/1905, and 2/10/1906, MC 268, Montana Power Company Predecessor Company Records, 1880–1947, Box 23/Folder 1, Montana Historical Society Archives, Helena (subsequent citations from this source will appear as "MPC Predecessor Company Records"); "Excavation of Channel," 11/5/1905, MC 268, Box 24/Folder 2, MPC Predecessor Company Records; Charles Kennedy, OH 266, and Hazel Karkanen, OH 392, Montanans at Work Oral History Project, Montana Historical Society, Helena (subsequent citations from this source will provide the OH number only).

7. Norman Smith, *A History of Dams* (London: Chaucer Press, 1971).

8. Ibid. Also see Daniel W. Mead, *Water Power Engineering: The Theory, Investigation and Development of Water Powers* (New York: McGraw-Hill, 1920).

9. For more examples of the size, capacity, and number of dams being built in the United States in the early twentieth century, see *Engineering News-Record*, vols. 52–62, 1904–1909.

10. Drawing "B," MSS 240, Montana Power Company Records, Series IV, DCb-1, Archives and Special Collections, Mansfield Library, University of Montana, Missoula.

11. Letters from Jerry Rourke to A. H. Wethey, 10/10/1905, 10/12/1905, 2/5/1906, 2/10/1906, MC 268, Box 23/Folder 1, MPC Predecessor Company Records; "Excavation of Channel," 11/5/1905, MC 268, Box 24/Folder 2, MPC Predecessor Company Records; Charles Kennedy, OH 266; and Hazel Karkanen, OH 392.

12. Letters from Slack to Wethey, 5/26/1907, 6/30/1907, 9/2/1907, 10/15/1907, 10/17/1907, 11/6/1907, MC 268, Box 23/Folder 2, MPC Predecessor Company Records.

13. Letters from Rourke to Wethey, 12/28/1905, 3/28/1906, 4/14/1906, 5/15/1906, 6/28–7/12/1906, MC 268, Box 23/Folder 1, MPC Predecessor Company Records; letters from Slack to Wethey, "George Slack, engineer Clark Dam, 2/19/1908, 5/1907–6/1908," MC 268, Box 23/Folder 2, MPC Predecessor Company Records.

14. Letters from Slack to Wethey, 4/6/1908, 4/20/1908, 5/6/1908, MC 268, Box 23/Folder 2, MPC Predecessor Company Records.

15. "Power Plant at Bonner Tested," *Daily Missoulian* (Missoula, Mont.), 1/10/1908. For similar accounts also see "Big Dam near Bonner Is Finally Completed," *Anaconda (Mont.) Standard*, 1/10/1908; "To Test Big Plant This Morning," *Daily Missoulian*, 1/7/1908; and "New Dam at Bonner Is Flooded," *Daily Missoulian*, 1/9/1908.

16. "Power Plant at Bonner Tested," *Daily Missoulian*, 1/10/1908. For similar accounts also see "Big Dam near Bonner Is Finally Completed," *Anaconda Standard*, 1/10/1908; "To Test Big Plant This Morning," *Daily Missoulian*, 1/7/1908; and "New Dam at Bonner Is Flooded," *Daily Missoulian*, 1/9/1908.

17. Letter from Charles Herbert McLeod to Charles A. Vail, MSS 581, Missoula Light and Water Company Records, Series II, Box 1, V.26, "Financial Correspondence, 1901–1906," Archives and Special Collections, Mansfield Library, University of Montana, Missoula; letter from Wethey to McLeod, MSS 001, Charles Herbert McLeod, Series I: Correspondence, 1885–1947, Archives and Special Collections, Mansfield Library, University of Montana, Missoula; "Work Soon to Begin on the Dam," *Daily Missoulian*, 9/15/1905. On Clark's entrepreneurial background see William Daniel Mangam, *Biography of W. A. Clark and His Tarnished Family* (Butte, Mont.: Old Butte Publishing, 2007); and C. B. Glasscock, *The War of the Copper Kings: Greed, Power, and Politics: The Billion-Dollar Battle for Butte, Montana, the Richest Hill on Earth* (1935; repr., Helena, Mont.: Riverbend, 2002).

18. Letters from Rourke to Wethey, 4/24/1906, 5/10/1906, 8/22/1906, 9/13/1906, MC 268, Box 23/Folder 1, MPC Predecessor Company Records; letters from Slack to Wethey, 1/23/1908, 1/26/1908, 5/11/1908, MC 268, Box 23/Folder 2, MPC Predecessor Company Records.

19. "Power Plant at Bonner Tested," *Daily Missoulian*, 1/10/1908.

20. Letters from Slack to Wethey, 3/20/1908, 4/27/1908, MC 268, Box 23/Folder 2, MPC Predecessor Company Records.

21. Letters from Slack to Wethey, 5/22/1908, 5/28/1908, 5/31/1908, MC 268, Box 23/Folder 2, MPC Predecessor Company Records.

22. "Rainfall During May Was Greatest in Years," *Stevensville (Mont.) Northwest Tribune*, 6/5/1908, "Rain Holds Reins and Reigns," *Stevensville Northwest Tribune*,

6/12/1908; "History of the Greatest Disaster That Missoula Has Yet Suffered," *Stevensville Northwest Tribune*, 6/19/1908; "Residents in Missoula Are Swept down River," *Anaconda Standard*, 6/7/1908; "Rain the Heaviest in Years," *Daily Missoulian*, 6/6/1908.

23. "Great Damage Is Wrought by the Flood Waters in Western Part of Montana," *Daily Missoulian*, 6/6/1908.

24. "History of the Greatest Disaster That Missoula Has Yet Suffered," *Stevensville Northwest Tribune*, 6/19/1908; "Great Damage Is Wrought by the Flood Waters in Western Part of Montana," *Daily Missoulian*, 6/6/1908; "Long Journey in the Mud," *Daily Missoulian*, 6/6/1908; "Houses Borne Away," *Daily Missoulian*, 6/7/1908.

25. "Flood Situation in State Reaches a Serious Stage," *Daily Missoulian*, 6/5/1908; "Great Damage Is Wrought by the Flood Waters in Western Part of Montana," *Daily Missoulian*, 6/6/1908; "Section of Higgins Avenue Bridge Collapses," 6/7/1908; "City in State of Alarm," *Daily Missoulian*, 6/7/1908.

26. "Crisis Now Passed at Bonner," "Firm as a Mountain Is the Dam," *Daily Missoulian*, 6/7/1908.

27. "Dam Is Solid As a Rock," *Daily Missoulian*, 6/8/1908; letters from Rourke to Wethey, 2/10/1906, 2/16/1906, MC 268, Box 23/Folder 1, MPC Predecessor Company Records; letter from M. Gerry to W. A. Clark, 3/20/1906, MC 268, Box 23/Folder 1, MPC Predecessor Company Records.

28. "Memorandum of Inspection and Investigation of Undercut Repairs to the Missoula-Bonner Dam, Made during the Summer and Fall of 1930," MSS 240, Montana Power Company Records, Series VII and Series XXXVII, Box 1, Folder 6, Archives and Special Collections, University of Montana.

29. Nielsen, "Milltown Dam History."

30. Steve Woodruff, "Leaks in Milltown Dam Concern Power Company," *Missoulian*, 10/25/1983.

31. No byline, "Lower Pond, Check Dam, Federal Agency Tells MPC," *Missoulian*, 10/27/1983; no byline, "Milltown Reservoir Lowered," *Missoulian*, 11/15/1983.

32. Steve Woodruff, "MPC Weighs Fate of Dam: Options Include Abandoning Milltown Structure," *Missoulian*, 11/18/83.

33. No byline, "Minimum Flows Sought for Clark Fork River," *Missoulian*, 11/23/1983.

34. Steve Woodruff, "Congressmen Call on EPA to Study Clark Fork River," *Missoulian*, 2/29/1984; Tim Palmer, *Endangered Rivers and the Conservation Movement* (Lanham, Md.: Rowman and Littlefield, 2004).

35. Editorial, "Government Responded Well with River Study," *Missoulian*, 3/12/1984.

36. Elizabeth D. Blum, *Love Canal Revisited: Race, Class, and Gender in Environmental Activism* (Lawrence: University Press of Kansas, 2008).

37. Steve Shirley, "River Study Set," *Missoulian*, 3/8/1984; Missoulian State Bureau, "New Image?" *Missoulian*, 3/10/1984; editorial, "Government Responded Well with River Study," *Missoulian*, 3/12/1984; Cathy Kradolfer, "Clark Fork Study Set: State, Industry Join on Detailed Basin Look," *Missoulian*, 5/1/1984.

38. Steve Shirley, "River Study Set," *Missoulian*, 3/8/1984; Missoulian State Bureau, "New Image?" *Missoulian*, 3/10/1984; editorial, "Government Responded Well with River Study," *Missoulian*, 3/12/1984.

39. Bob Anez, "More Than 1,000 Clark Fork Fish Killed," *Missoulian*, 8/4/1984; no byline, "Clark Fork Fish Kill Blamed on High Levels of Copper," *Missoulian*, 8/9/1984.

40. For Butte and Anaconda mining and smelting history, see Donald MacMillan, *Smoke Wars: Anaconda Copper, Montana Air Pollution, and the Courts, 1890–1920* (Helena: Montana Historical Society Press, 2000); Stiller, *Wounding the West*; Malone, Roeder, and Lang, *Montana*; and Michael Punke, *Fire and Brimstone: The North Butte Mining Disaster of 1917* (New York: Hyperion, 2006).

41. Carolyn Johns and Johnnie Moore, *Heavy Metals in Bottom Sediments of Clark Fork River Reservoirs*, report to EPA, 7/11/1985, reel 601; Johnnie Moore, *Source of Metal Contamination in Milltown Reservoir, Montana*, report to EPA, 9/1985, reel 146, Milltown Administrative State Site Records; letters from George Slack to A. H. Wethey, 1908, MC 268, Box 23/Folder 2, MPC Predecessor Company Records. Also see Malone, Roeder, and Land, *Montana*; Malone, *Battle for Butte*; and Stiller, *Wounding the West*.

42. Fred Quivik, "Smoke and Tailings: An Environmental History of Copper Smelting Technologies in Montana, 1880–1930" (PhD diss., University of Pennsylvania, 1998), v, 177–94.

43. Ibid., 195–203, 280–86, 296.

44. Steve Woodruff, "Biologists Say River Study Is Only a First Step," *Missoulian*, 10/31/1984.

45. Gordon Gregory, "Watchers: Clark Fork Poor Habitat," *Missoulian*, 9/17/84.

46. Steve Woodruff, "MPC to Fix Milltown Dam," *Missoulian*, 12/12/1984.

47. Ibid.

48. Editorial, "Good News about the Milltown Dam," *Missoulian*, 12/16/1984.

49. Steve Woodruff, "MPC Plan for Dam Would Cost $11.5 Million," *Missoulian*, 6/15/1985.

50. Steve Woodruff, "Clark Fork Coalition Meeting to Focus on MPC's Milltown Dam," *Missoulian*, 1/8/1985.

51. Daniel Kemmis, "Riverfront Efforts: Good Work, Missoula!" *Missoulian*, 8/7/1985.

52. Patricia Sullivan, "Working on a 'Chain Gang': Volunteers Sweat to Clean Clark Fork's South Bank," *Missoulian*, 7/22/1985.

53. Steve Woodruff, "EPA Considers Clark Fork as Possible Superfund Site," *Missoulian*, 6/20/1985.

54. Editorial, "Welcome News Flows for the Clark Fork," *Missoulian*, 7/31/1985.

55. Jonathan Friendly, "President Asks 5-Year Extension of Toxic Waste Cleanup Program," *New York Times*, 2/23/1985.

56. "Superfund: CERCLA Overview," U.S. Environmental Protection Agency, http://www.epa.gov/superfund/policy/cercla.htm.

57. Gary Klott, "Superfund Tax Stirs Debate," *New York Times*, 10/26/85; editorial, "Tough Superfund Bill Should Be Passed," *Missoulian*, 7/25/1985.

58. Philip Shabecoff, "Lack of Money Might Halt Toxic Cleanup Soon, U.S. Officials Say," *New York Times*, 1/26/86.

59. State News Service, "'Infighting' Is Said to Delay Superfund," *New York Times*, 6/22/86; editorial, "The Superfund: Locked in a Lovefest," *New York Times*, 7/19/86; Philip Shabecoff, "Head of Agency Warns Congress on Toxic Waste Cleanup Projects," *New York Times*, 9/24/86.

60. Steve Woodruff, "Life at EPA 'Was Very Chaotic,'" *Missoulian*, 3/24/1985.

61. Ibid. Also see Hays, *Beauty, Health, and Permanence*, 491–526; and Norman J. Vig and Michael E. Kraft, *Environmental Policy in the 1980s: Reagan's New Agenda* (Washington, D.C.: Congressional Quarterly, 1984).

62. "Federal Register Notices for NPL Updates," National Priorities List, U.S. Environmental Protection Agency, http://www.epa.gov/superfund/sites/npl/frlist.htm.

63. Tom Cook (Missoulian State Bureau), "Toxic Waste: Superfund Bill Draws Criticism," *Missoulian*, 12/6/1985.

64. Max Baucus in Environment and Natural Resources Policy Division, *A Legislative History of the Superfund Amendment and Reauthorization Act of 1986 (Public Law 99–499)* (Washington, D.C.: U.S. Government Printing Office, 1990), 2:1068–71; Tom Cook (Missoulian State Bureau), "Activists Bemoan Superfund Measure," *Missoulian*, 10/25/1985.

65. Tom Cook (Missoulian State Bureau), "Two Governors Assail Superfund Cleanup Bill," *Missoulian*, 11/3/1985.

66. Environment and Natural Resource Policy Division, *Legislative History*, vol. 1; or see CERCLA section 9625, "42 U.S. Code § 9625," Cornell University Law School, accessed 2/22/2012, http://www.law.cornell.edu/uscode/text/42/9625; CERCLA section 9617, "42 U.S. Code § 9617," Cornell University Law School, accessed 11/12/20104, http://www.law.cornell.edu/uscode/text/42/9617.

67. Max Baucus in "Superfund Controversy," *Congressional Digest*, June–July 1986, 172.

68. Tom Cook, "Experts Optimistic about Clark Fork," *Missoulian*, 6/12/1985; John Stromnes, "EPA Ready to Widen Study of Clark Fork," *Missoulian*, 7/25/1985.

69. Steve Woodruff, "Flooding Eases; Dam Faces Repairs," *Missoulian*, 2/27/1986; and "Flood Damage to Milltown Dam Forces MPC to Plan Repairs," *Missoulian*, 2/28/1985.

70. Steve Woodruff, "Schwinden: Clark Fork Cleanup Should Progress," *Missoulian*, 3/23/1986.

71. Steve Woodruff, "Montana Power Warns of Dam Danger," *Missoulian*, 3/26/1986.

72. Steve Woodruff, "Conditions Attached to Dam Reconstruction," *Missoulian*, 4/12/1986.

73. Steve Woodruff, "Milltown Dam Plans May Alter: MPC May Receive OK for Only Part of Proposal for Rebuilding Structure," *Missoulian*, 5/2/1986; editorial, "Let's

Keep the River 'Naturally Inviting,'" *Missoulian*, 5/9/1986; Larry Howell, "Milltown Dam Work Receives Conditional OK," *Missoulian*, 6/17/1986; editorial, "Go-Ahead at the Milltown Dam," *Missoulian*, 6/22/1986.

Chapter 3

1. Federal Energy Regulatory Commission, "Environmental Assessment: Milltown Hydroelectric Project," *Missoulian*, 6/30/1986; Theresa Johnson, "Redoing a Dam," *Missoulian*, 8/12/86; Michael Moore, "Dam Work Flowing Smoothly," *Missoulian*, 9/19/86; opinion, "Menace to River Still Caged," *Missoulian*, 9/23/86; Michael Moore, "Ready for Runoff," *Missoulian*, 3/27/87.

2. Peter Nielsen, e-mail message to author, 9/8/2011; Peter Nielsen (CFC) to U.S. Army Corps of Engineers, 9/29/1987, reel 150, Milltown Administrative State Site Records; Kent Gilbreath, "Industry's Environmental Attitudes," *EPA Journal*, July/August 1988, 19; Palmer, *Endangered Rivers*.

3. See Rothman, *Greening of a Nation*; Kirkpatrick Sale, *The Green Revolution: The American Environmental Movement, 1962–1992* (New York: Hill and Wang, 1993); and Wellock, *Preserving the Nation*.

4. Hays, *Beauty, Health, and Permanence*, 491, 505.

5. Ibid., 529.

6. Letter from John F. Wardell to Frank Pickett, 11/29/1984, reel 169, Milltown Administrative State Site Records. For specifics on Superfund law, see Sections 104, 106, and 107 of CERCLA. "Enforcement: Superfund Enforcement," U.S. Environmental Protection Agency, http://www.epa.gov/compliance/cleanup/superfund/orders.html.

7. Michael E. Zimmerman to Vic R. Anderson, 12/10/1984; MPC, "Summary: MPC Milltown Hydroelectric Project Engineering Evaluation"; MPC, "Water Resources Impacts Resulting from Proposed Rehabilitation of the Milltown Hydroelectric Project," 5/3/1985; all on reel 169, Milltown Administrative State Site Records.

8. "United States of America before the Federal Energy Regulatory Commission," Project No. 2543, 3/27/1986, reel 148; letter from Stan Bradshaw (FWP) to Kenneth F. Plumb (FERC), 3/28/1986, reel 150; "Comments and Recommendations: Submitted by the MDNRC as a Supplement to Paragraph B of Its Intervention of the Milltown Dam Amendment," 4/6/1986, reel 169; "Comments of the Clark Fork Coalition to the Federal Energy Regulatory Commission," 5/19/1986, reel 169, Milltown Administrative State Site Records.

9. Letter from John F. Wardell to Robert J. Labrie, MPC senior vice president, 3/5/1986; news release, Montana Department of Health and Environmental Sciences, 5/30/1986; "EPA Comments Following Review of MPC's FERC Application," n.d.; Richard Hunt to Labrie, 6/16/1986; all on reel 169, Milltown Administrative State Site Records.

10. Jane L. Bloom, "Superfund: The Next Five Years," *Environment* 27, no. 4 (May 1985); Associated Press (Butte, Mont.), "Knell Sounds for Superfund Sites' Cleanup," *Missoulian*, 6/4/1986; opinion, "Superfund Veto Risky for Montanans," *Missoulian*, 10/15/86. Also see Revesz and Steward, *Analyzing Superfund*; Vig and Kraft,

Environmental Policy; and Ronald Reagan, *The Reagan Diaries*, ed. Douglas Brinkley (New York: HarperCollins, 2007), 131–32, 252–53, 445.

11. EPA, "Administrative Order of Consent: Milltown Reservoir Sediments Site," reel 241, Milltown Administrative State Site Records.

12. Revesz and Steward, *Analyzing Superfund*, 3–6; "Superfund: SARA Overview," U.S. Environmental Protection Agency, http://www.epa.gov/superfund/policy /sara.htm; U.S. EPA, *Milltown Reservoir Sediments Operable Unit: Record of Decision*, 2/2004, 2–7, available at U.S. EPA Region 8 office, Helena, Mont.

13. Letter from Benjamin S. Bilus to Stephen H. Foster, Esquire, 7/28/1986, reel 144, Milltown Administrative State Site Records.

14. EPA, "Initial Site Evaluation," Summer 1988, reel 148; EPA, "Administrative Order on Consent," 2/14/1990, reel 241; EPA, "Chain-of-Title for the Milltown Reservoir," n.d., reel 144, Milltown Administrative State Site Records; Environment and Natural Resources Policy Division, *Legislative History*, 1:45.

15. Letter from EPA to John Drynan, director Montana DHES, 8/27/1986, reel 148, Milltown Administrative State Site Records; "Superfund: SARA Overview," U.S. Environmental Protection Agency, http://www.epa.gov/superfund/policy/sara.htm.

16. Letter from James W. Flynn (FWP) to Kenneth Plumb (FERC), 2/24/1987 reel 150; letters from EPA to FERC, 3/24/1986, 5/23/1986, 2/10/1987, 3/17/1987, reel 169; letter from CFC to FERC, 3/25/1987, reel 169, Milltown Administrative State Site Records.

17. Letter from Wardell (EPA) to Plumb (FERC), 9/11/1987; letter from Wardell to Morton (U.S. Army Corps of Engineers), 9/11/1987; letter from Nielsen (CFC) to U.S. Army Corps of Engineers, 9/29/1987; letter from Nielsen (CFC) to Kenneth Plumb (FERC), 10/8/1987; all on reel 150, Milltown Administrative State Site Records.

18. EPA memorandum, "Milltown Dam Rehab Meeting," 10/20/1987, reel 150, Milltown Administrative State Site Records.

19. Letter from G. Howard Van Noy (MPC) to Kenneth Plumb (FERC), 6/4/1987, reel 148, Milltown Administrative State Site Records.

20. Clark Fork Coalition, "Milltown Rebuild Back on the Drawing Board," *Currents*, October 1987, reel 148; and "Milltown Dam Rebuild Sparks Renewed Interest," *Currents*, December 1987, reel 148; meeting notes, "Milltown Dam Rehab Phase II," 11/12/1987, reel 150; letter from EPA to MPC, 11/17/1987, reel 150; EPA memorandum, "Milltown Dam Rehabilitation Meeting, 11/18/1987, reel 150; letter from MPC to FWP, 11/19/1987, reel 150, Milltown Administrative State Site Records.

21. Dick Manning, "FERC Paves Way for Rebuilding of Milltown Dam," *Missoulian*, 3/3/1988; editorial, "A Reasonable Decision," *Missoulian*, 3/7/1988.

22. Dick Manning, "MPC Appeals Milltown Dam Order," *Missoulian*, 3/17/1988; Dick Manning, "MPC Drains Milltown Reservoir for Project," *Missoulian*, 7/11/1988.

23. Record of telephone communication, Jim Duncan to Ken Wallace (EPA), 7/5/1988, reel 169; attachment to phone log (EPA), "Significance of Sole Source Aquifer Designation," reel 169, Milltown Administrative State Site Records; editorial, "Cleaner Clark Fork on Tap," *Missoulian*, 7/19/1988.

24. Dick Manning, "Project Delay Darkens Clark Fork," *Missoulian*, 7/22/1988; Dick Manning, "Spill Poses No Human Risk," *Missoulian*, 7/26/88; no byline, "Milltown Work Begins," *Missoulian*, 7/29/1988.

25. Letter from CFC to EPA, 8/1/1988, reel 144, Milltown Administrative State Site Records.

26. Letter from Phil Herzog (DHES) to Phil Tourangeau (CFC), 8/10/1988, reel 144, Milltown Administrative State Site Records.

27. Dick Manning, "EPA Wants Milltown Work Hastened," *Missoulian*, 8/17/1988.

28. Revesz and Steward, *Analyzing Superfund*, 3–6, also chap. 6, "Evaluating the Impact of Alternative Superfund Financing Schemes"; EPA, "Initial Site Evaluation," 1988, reel 148, Milltown Administrative State Site Records.

29. Dick Manning, "MPC, State Make Deal," *Missoulian*, 10/20/1988; Dick Manning, "FERC Refuses to Extend License for Milltown Dam," *Missoulian*, 1/10/1989.

30. Letter from landowners (Schmaus, Mee, and Morcross) to MPC, 9/8/1988, reel 144, Milltown Administrative State Site Records.

31. Letter from Ken Wallace (EPA) to Tina Schmaus and Gaynelle Mee, 11/8/1988, reel 144; letter from Schmaus and Mee to EPA, 11/18/1988, reel 144, Milltown Administrative State Site Records.

32. Letter from FERC to MPC, 12/7/1988, reel 148; letter from MPC to FERC, 12/27/1988, reel 148; letter from Mee to FERC, 3/29/1989, reel 603, Milltown Administrative State Site Records; James J. Florio, "Public Role in Toxic Cleanup," *New York Times*, 3/15/1987.

33. Dick Manning, "The Matter of Milltown Dam," *Missoulian*, 6/1/1989; Dick Manning, "MPC Resolves Grievances at Bonner," *Missoulian*, 7/11/1989; letter from FERC to Mee, 6/28/1989, reel 148, Milltown Administrative State Site Records.

34. Letter from Robert Dent (ARCO) to Ken Wallace (EPA), 6/17/1988, reel 144; ARCO, "1989 First Quarter Report," reel 241, Milltown Administrative State Site Records; Philip Shabecoff, "Toxic Cleanup Plan Moves Slowly amid Criticism from Two Fronts," *New York Times*, 1/10/1988; William K. Stevens, "Saga of a Waste Cleanup: 12 Years and Counting," *New York Times*, 10/28/1988; Associated Press, "Toxic Site Cleanup Reported Lagging," *New York Times*, 9/10/1989; Thomas S. Clavin, "Toxic-Site Cleanups Inching Along," *New York Times*, 12/16/1990; Barnaby J. Feder, "In the Clutches of the Superfund Mess," *New York Times*, 6/16/1991.

35. "Atlantic Richfield Company History," FundingUniverse, http://www.fundinguniverse.com/company-histories/atlantic-richfield-company-company-history.html; Edwin Dobbs, "Pennies from Hell," *Harper's Magazine*, October 1996.

36. EPA, "Administrative Order on Consent, Milltown Reservoir Sediments Site," 2/14/1990, reel 241, Milltown Administrative State Site Records. For specifics on the decentralization of EPA authority, also see "Executive Order 12580—Superfund Implementation," National Archives, Federal Register, accessed 11/15/2014, http://www.archives.gov/federal-register/codification/executive-order/12580.html; U.S. EPA, "Superfund Internal Delegations of Authority," 4/1/1990, National Service Center for Environmental Publications, accessed 11/15/2014, http://nepis.epa.gov; and Ross Ettlin, "Facts to Reflect On," *EPA Journal*, September/October 1990, 29.

37. EPA press release, 6/23/1989; letter from ARCO to Wardell (EPA), 11/29/1989; EPA, "Administrative Order on Consent, Milltown Reservoir Sediments Site," 2/14/1990, all on reel 241, Milltown Administrative State Site Records; "Enforcement: Superfund Enforcement," U.S. Environmental Protection Agency, http://www.epa.gov/compliance/cleanup/superfund/neg-type.html#aas.

38. EPA minutes, "Citizen's Advisory Group at Milltown Superfund Site," 7/24/1989, reel 465; letter from Ken Wallace (EPA) to Phil Tourangeau (CFC), 9/21/1989, reel 241, Milltown Administrative State Site Records.

39. EPA press release, 2/8/1990, reel 241, Milltown Administrative State Site Records.

40. MPC to FERC, "Application of the Montana Power Company for Amendment of License to Extend the Date for Termination of the License," 4/10/89, reel 148, Milltown Administrative State Site Records.

41. William K. Reilly, A *Management Review of the Superfund Program* (Washington, D.C.: U.S. Environmental Protection Agency, 1989), ii.

Chapter 4

1. David Quammen, "Alias, Benowitz Shoe Repair: Heavy Metal, High Water, and a Man in a Neoprene Suit," *Outside*, December 1983; George Ochenski, "Curing the Clark Fork: Poisons from a Century of Mining Have Given Montana Its Own Love Canal," reel 397, Milltown Administrative State Site Records.

2. Probst et al., *Footing the Bill.*

3. Bill Wilke, "Government Sues ARCO, Wrecking Firm," *Missoulian*, 6/27/1989; "42 U.S. Code § 9601—Definitions," Cornell University Law School, accessed 11/12/2014, http://www.law.cornell.edu/uscode/text/42/9601; Probst et al., *Footing the Bill.*

4. Editorial, "Arco Offer Is Only a Start," *Missoulian*, 8/18/1989.

5. See "1999 Consent Decree between State and ARCO," Montana Department of Justice, https://dojmt.gov/lands/lawsuit-history-and-setttlements/#montanaarco1999.

6. John Stromnes, "Clark Fork Fish Kill: Old Toxins Wipe Out Trout along 18 Miles of River," *Missoulian*, 7/14/1989; John Stromnes, "Arco Says High Water Delayed Dam Project," *Missoulian*, 7/15/1989. Also see Steve Woodruff, "Clark Fork Killer: Biologist Says He's Cracked the Case of the Dying Fish," *Missoulian*, 7/28/1987; and Sherry Devlin, "Legacy of Pollution: State Report Documents Clark Fork River Fish Kill," *Missoulian*, 6/8/1993.

7. Letter from Johnnie Moore (UM) to Donald Pizzini (DHES), 11/8/1989, reel 565; letters from Pizzini to Vicki Watson (UM) and Peter Nielsen (CFC), 11/9/1989, reel 602; letter from Pizzini to Moore, 11/20/1989, reel 602; letter from Nielsen to John Wardell (EPA), 5/8/1990, reel 565; letter from Cindy Nelson (Deer Lodge Valley Conservation District) to DHES, 5/14/1990, reel 565, Milltown Administrative State Site Records. Also Associated Press (Butte, Mont.), "'Chronic' Fish Mortality Affects Clark Fork," *Missoulian*, 10/15/1989.

8. Editorial, "Arco Offer Is Only a Start," *Missoulian*, 8/18/1989; Sherry Devlin, "State Isn't Ready for ARCO, Clark Fork Coalition Says," *Missoulian*, 9/14/1990;

"42 U.S. Code § 9607(f)(1)—Liability," Cornell University Law School, accessed 11/12/2014, http://www.law.cornell.edu/uscode/text/42/9607.

9. Letter from Wardell to Nielsen, 6/27/1989, reel 241, Milltown Administrative State Site Records.

10. David Wann (EPA, Region 8), "On the Firing Line: The Challenge of Environmental Risk in Region 8," *EPA Journal*, November 1987, 30; Lee M. Thomas, "The Challenge of Community Involvement," *EPA Journal*, December 1985, 2; Daphne Gemmill and Edwin Berk, "Community Involvement in Superfund: The Results," *EPA Journal*, December 1985, 7–9.

11. U.S. EPA news release, 4/5/1991, reel 397; Tina Reinicke-Schmaus, "A Crash Course in Superfund," *MTAC News*, February 1993, reel 601, Milltown Administrative State Site Records. Also see "42 U.S. Code § 9617—Public Participation," Cornell University Law School, accessed 11/12/2014, http://www.law.cornell.edu/uscode/text/42/9617.

12. U.S. EPA, Region 8, "Administrative Order on Consent: Milltown Reservoir Sediments Site," reel 399; letter from Barry Dutton (Missoula Water Quality Advisory Group) to DalSoglio (EPA), 3/28/1991, reel 397; letter from Missoula City-County Health Dept. to DalSoglio, 3/30/1991, reel 397; letter from Joe DosSantos (Montana chapter of American Fisheries Society) to DalSoglio, 3/29/1990, reel 397, Milltown Administrative State Site Records.

13. Bob Narus, "Waste Sites: How Clean Will They Be?," *New York Times*, 10/26/1986; "Progress and Challenges: Looking at EPA Today," *EPA Journal*, September/October 1990, 24.

14. Bob Narus, "Waste Sites: How Clean Will They Be?," *New York Times*, 10/26/1986; James J. Florio, "Public Role in Toxic Cleanup," *New York Times*, 3/15/1987; editorial, "Not So Super Superfund," *New York Times*, 2/7/1994. Also see "42 U.S. Code § 9621—Cleanup Standards," Cornell University Law School, accessed 11/12/2014, http://www.law.cornell.edu/uscode/text/42/9621; "Progress and Challenges: Looking at EPA Today," *EPA Journal*, September/October 1990, 24.

15. Letter from George Ochenski to DalSoglio, 3/31/1991, reel 397, Milltown Administrative State Site Records, emphasis in original.

16. Probst et al., *Footing the Bill.*

17. Henry L. Longest, "Superfund Wasn't Meant to Be a Cure-All," *New York Times*, 8/18/1991; Keith Schneiders, "E.P.A. Urged to Ease Rules on Cleanup of Toxic Waste," *New York Times*, 2/25/1992.

18. Keith Schneiders, "New Breed of Ecologist to Lead E.P.A.," *New York Times*, 12/17/1992.

19. Keith Schneiders, "New View Calls Environmental Policy Misguided," *New York Times*, 3/21/1993.

20. John H. Cushman, Jr., "Overhaul of Toxic-Waste Cleanup Program Passes First Test," *New York Times*, 5/12/1994; John H. Cushman, Jr., "Demoralized E.P.A. Works to Set Goals," *New York Times*, 9/24/1995; John H. Cushman, Jr., "Republicans Back Off Plan to Ease Polluters' Liability," *New York Times*, 9/29/1995; Bill Line, "What

Voters Say about the Environment Today: Poll Shows Widespread Support for Green Causes," *EPA Journal*, Winter 1995.

21. Probst et al., *Footing the Bill*, 17; John H. Cushman, Jr., "Moderates Soften G.O.P. Agenda on Environment," *New York Times*, 10/24/1995; John J. Florio, "Gingrich Revolution Gives Polluters a Rebate on Cleanup Costs," *New York Times*, 1/4/1996.

22. John H. Cushman, Jr., "Program to Clean Toxic Waste is Left in Turmoil," *New York Times*, 1/15/1996; Andy Newman, "Cleanup of Toxic Waste Sites Falls Victim to the Federal Budget Tangle," *New York Times*, 1/21/1996; John H. Cushman, Jr., "G.O.P. Backing Off from Tough Stand over Environment," *New York Times*, 1/26/1996; Alison Mitchell, "Clinton Asks Tax Breaks for Toxic-Waste Cleanup," *New York Times*, 3/12/1996.

23. John H. Cushman, Jr. "Gingrich, Like the President, Calls for a 'New Environmentalism,'" *New York Times*, 4/25/1996; Jerry Gray, "Both Congress and Clinton Find Cause for Cheer in the Final Budget Deal," *New York Times*, 4/26/1996.

24. Letter from Phil Tourganeau to DalSoglio, 4/1/1991; letter from Peter Nielsen to DalSoglio, 4/1/1991; letter from Frank Rives to DalSoglio, 3/28/1991; letter from MPC to DalSoglio, all on reel 397, Milltown Administrative State Site Records.

25. Erick W. Bretthauer and Peter R. Jutro, "Science and the Regulatory Process," *EPA Journal*, March 1988, 13.

26. U.S. Fish and Wildlife Service (Montana Office), *Milltown Reservoir Sediments Site Endangerment Assessment Wildlife Survey: Final Draft*, 6/1991, reel 399; David C. Wilborn and Clarence A. Callahan, "On-Site Earthworm (*Eisenia foetida*) Test for the Evaluation of Toxicity and Uptake of Contaminants at the Milltown Superfund Site," 2/1990, reel 645; Ken Knudsen, "Preliminary Estimates for Trout Densities in the Middle Clark Fork River if Milltown Dam and Reservoir Were Not Present," 4/24/1992, reel 684; U.S. EPA and Environmental Toxicology International, Inc., "Minutes of Meetings with Wetlands Work Group, Public Health Work Group, and Fisheries Work Group," 2/10/1992, reel 465; EPA and Montana DHES, "Milltown Reservoir: Superfund Site Report," 7/1992, reel 565; University of Wyoming, "Milltown Endangerment Assessment Project: Effects of Metal-Contaminated Sediment, Water, and Diet on Aquatic Organisms," 9/4/1992, reel 564; letter from ARCO to DalSoglio (EPA), 10/19/1992, reel 601; McGuire Consulting, "Clark Fork River Macroinvertebrate Community Biointegrity: 1993 Assessment," 8/1994, reel 768, Milltown Administrative State Site Records.

27. MPC, "Milltown Fisheries Protection, Mitigation, and Enhancement Plan," 10/29/1992, reel 601, Milltown Administrative State Site Records.

28. U.S. EPA and Environmental Toxicology International, Inc., "Minutes of Meetings with Wetlands Work Group, Public Health Work Group, and Fisheries Work Group," 2/10/1992, reel 465; Environmental Toxicology International, Inc. for EPA, "Baseline Human Health Risk Assessment: Milltown Reservoir Operable Unit," reel 645, and "Exposure Survey Results: Milltown Reservoir Sediments Site Public Health Evaluation," 7/1993, reel 712, Milltown Administrative State Site Records.

29. Letter from Barbara Bennetts to Steve Blodgett, 10/31/1993, reel 684; letter from Montana DHES to DalSoglio (EPA), 11/3/1993, reel 712; letter from EPA to Montana DHES, 12/27/1993, reel 684; letter from MCCHD to Montana DHES, 1/5/1994, reel 885; letter from Julie DalSoglio (EPA) to Barbara Bennetts, 2/1/1994, reel 885; "ARCO Comments on the Baseline Human Health Risk Assessment Milltown Reservoir Operable Unit, Milltown Reservoir Sediments Site," 12/1/1993, reel 684, Milltown Administrative State Site Records.

30. Letter from Jane Heath (EPA) to multiple EPA recipients, "RE: Meeting with ATSDR about Pancreatic Cancer Study and other activities," 5/10/1994, reel 885; Todd Damrow (ATSDR), handwritten notes, 7/17/1995, reel 885; letter from Dal-Soglio to Bennetts, 8/1/1995, reel 821, Milltown Administrative State Site Records.

31. EPA and Montana DHES, "Milltown Reservoir: Superfund Site Report," July 1992, reel 565; notes from telephone conversation between Robert Fox (EPA) and Sandy Stash (ARCO), 12/11/1992, reel 602, Milltown Administrative State Site Records.

32. Editorial, "Progress and Challenges: Looking at EPA Today"; and Ross Ettlin, "Facts to Reflect On," *EPA Journal*, September/October 1990; Julie DalSoglio, "Clark Fork Coordinating Forum Minutes," 1/21/1993, reel 601, Milltown Administrative State Site Records.

33. Keith Schneider, "Old Peril of Abandoned Mines Stirs Calls for Action in West," *New York Times*, 4/27/1993.

34. William K. Reilly, "Why I Propose a National Debate on Risk," *EPA Journal* 17 (March/April 1991): 2–3; D. Warner North, "Forum Two—The Science," *EPA Journal* 17 (March/April 1991); U.S. Environmental Protection Agency, "Profiles in Risk Assessment: New Science, New Contexts," *EPA Journal* 19, no. 1 (January/February/March 1993).

35. Letter from C. B. Pearson (CFC) to Wardell, 11/19/1993, reel 684; State of Montana Natural Resource Damage Program, "Assessment Plan: Part I, Clark Fork River Basin NPL Sites, Montana," January 1992, reel 821; letters from EPA to Bruce Farling (CFC) and Peter Nielsen (MCCHD), 5/20/1992, reel 583; letter from Kenneth Brett to Wardell, 12/3/1993, reel 684; letter from Ellen Leahy (MCCHD) to Wardell, 12/23/1993, reel 684; letter from Kevin Kirley (MDHES) to DalSoglio, 12/27/1993, reel 684; Steve Blodgett, "EPA Decides Not to Defer at Milltown," *MTAC News*, March 1994, reel 821, Milltown Administrative State Site Records.

36. Letter from C. Floyd George (ARCO) to Jack McGraw (EPA), 11/24/1993, reel 684, Milltown Administrative State Site Records.

37. Joshua Lipson, "Assessing Impacts to Aquatic Resources of the Clark Fork River, Montana," *Inside Hagler Bailly*, 10/1993, reel 768, Milltown Administrative State Site Records.

38. Titan Environmental Corporation, "Milltown Reservoir Sediments Operable Unit: Screening of Alternatives Summary Report," 11/1994, reel 768, Milltown Administrative State Site Records.

39. ARCO, "Milltown Reservoir Sediments Operable Unit: Final Remedial Investigation Report," 2/1995, reel 767, Milltown Administrative State Site Records.

40. Brian Sherry (MTAC), "KUFM Radio Commentary," 10/20/1994, 11/17/1994, 12/15/1994; letter from John Wardell, 2/15/1995, all on reel 768, Milltown Administrative State Site Records.

41. "'95 Legislative Digest—House Oks Suit Money," *Missoulian*, 3/2/1995.

42. Missoula City-County Health Department, "A Resolution Concerning the Cleanup of the Milltown Superfund Site," 2/16/1995, reel 768; letter from Jeffrey J. Smith to Rob Collins (NRDLP), 2/19/1995, reel 821, Milltown Administrative State Site Records.

43. Letter from Peter Nielsen (MCCHD) to Russ Forba (EPA), 3/10/1995, reel 768, Milltown Administrative State Site Records.

44. Letter from Geoffrey Smith (CFC) to Rob Collins (NRDLP), 3/22/1995, reel 821; letter from Brian Sherry (MTAC) to NRDLP, 3/23/1995, reel 821, Milltown Administrative State Site Records.

45. Letter from Peter Nielsen (MCCHD) to Rob Collins (NRDLP), 3/23/1995, reel 821, Milltown Administrative State Site Records, emphasis in original; Associated Press (Helena, Mont.), "'We Will Fight to the Very End,'" *Missoulian*, 10/31/1995; Company News, "ARCO Raises Cleanup Estimates by $1 Billion," *New York Times*, 3/4/1994.

46. Letter from Russ Forba (EPA) to Bill Bluck (CH2M), 4/12/95; EPA and MDHES, "Comments: Milltown FS Screening Report Addendum," 4/13/1995; letter from ARCO to EPA and MDHES, 5/30/1995; letter from Kirley (Montana Department of Environmental Quality) to ARCO, 7/20/1995, all on reel 821, Milltown Administrative State Site Records.

47. Letter from Kevin Kirley (MDHES) to Gary Fischer (MDNRC), 6/28/1995, reel 821, Milltown Administrative State Site Records.

48. EPA, "Public Notice," 12/1995, reel 887; ARCO, "Milltown Reservoir Sediments Site: Draft Feasibility Study Report," 4/1996, reel 932, Milltown Administrative State Site Records.

Chapter 5

1. Maclean, *A River Runs through It*; Edwin McDowell, "Publishing: Pulitzer Controversies," *New York Times*, 5/11/1984; Daryl Gadbow, "Reclaiming a River," *Missoulian*, 5/17/1990.

2. Sherry Devlin, "Babbitt: Listing Bull Trout Won't Hurt Fishing," *Missoulian*, 6/6/1998; Susan Gallagher, "Lawmakers Square Off over Bull Trout Protection," *Missoulian*, 3/5/1999; Paul Larmer, "Bull Trout History," *High Country News*, 9/15/1995; "Key Dates and Events in the History of the Bull Trout Listing," Alliance for the Wild Rockies, http://www.wildrockiesalliance.org/issues/bulltrout/history/bull trout_chronology.html.

3. U.S. Fish and Wildlife Service, Region 1, *Bull Trout* (Salvelinus confluentus) *Draft Recovery Plan*, 10/2002, reel 1523; CH2MHill, prepared for EPA, "Biological Assessment of the Milltown Reservoir Sediments Operable Unit," 4/2003, reel 1572, Milltown Administrative State Site Records; G. D. Holton and H. E. Johnson, *A Field Guide to Montana Fishes*, 3rd ed. (Helena: Montana Fish, Wildlife and Parks, 2003);

J. B. Dunham and B. E. Rieman, "Metapopulation Structure of Bull Trout: Influences of Physical, Biotic, and Geometrical Landscape Characteristics," *Ecological Applications* 9, no. 2 (May 1999): 642–55; Tim Swanberg, "The Movement and Habitat Use of Fluvial Bull Trout in the Upper Clark Fork River Drainage" (master's thesis, University of Montana, 1996).

4. Peter Nielsen, interview with author, MCCHD office, 10/14/2011; Mea Andrews, "Flood Watchers Still Wary," *Missoulian*, 2/11/1996.

5. Mea Andrews, "Flood Watchers Still Wary," *Missoulian*, 2/11/1996; Mea Andrews, "Statewide, the Worst of It's Over," *Missoulian*, 2/11/1996; Mick Holien, "Still Slamming, Jamming," *Missoulian*, 2/12/1996.

6. John Stromnes, "On Guard," *Missoulian*, 2/13/1996; Michael Moore, "It's Not Over Yet," *Missoulian*, 2/15/1996.

7. Sherry Devlin, "Milltown Drawdown Fouls River," *Missoulian*, 3/1/1996; Peter Nielsen, interview with author, 9/17/2009.

8. Sherry Devlin, "Major Fish Kill Feared in Clark Fork," *Missoulian*, 3/2/1996.

9. Peter Nielsen, interview with author, 9/17/2009; editorial, "Milltown Toxins Must Go," *Missoulian*, 3/3/1996.

10. MPC, "Milltown Dam Probable Maximum Flood," 6/10/1991, reel 601; in-house letter by CH2M, 4/22/1992, reel 601; EPA, "Task V, Review Literature on Dam-Breaks and Sediment Releases following Dam-Breaks," 5/1/1992, reel 564, Milltown Administrative State Site Records.

11. Missoula City-County Health Department, "Water Quality Sampling Associated with Ice Scouring and Flooding, Milltown Reservoir and Clark Fork River," 4/1/1996, reel 932, Milltown Administrative State Site Records.

12. Letter from Westslope chapter Trout Unlimited to Carol Browner (EPA), 3/6/1996, reel 932, Milltown Administrative State Site Records.

13. Letter from Westslope chapter Trout Unlimited to Carol Browner (EPA), 3/6/1996; Missoula Board of County Commissioners and Missoula City-County Board of Health, "Resolution to the United States Environmental Protection Agency concerning Contamination of Ground and Surface Waters below the Milltown Dam," 3/21/1996; letter from citizen group to John Wardell (EPA), 4/3/1996; letter from Erik Ringelberg (Missoula City-County Water Quality Advisory Council), 4/17/1996, all on reel 932, Milltown Administrative State Site Records; Sherry Devlin, "No-Pollution Parity," *Missoulian*, 3/15/1996.

14. Letter from John Wardell (EPA) to Robert Whalen (Westslope chapter Trout Unlimited), 3/28/1996, reel 932, Milltown Administrative State Site Records; Sherry Devlin, "EPA Postpones Decision on Milltown Cleanup," *Missoulian*, 5/1/1996.

15. Shreekant Gupta, George Van Houtven, and Maureen Cropper, "Paying for Permanence: An Economic Analysis of EPA's Cleanup Decisions at Superfund Sites," *Rand Journal of Economics* 27, no. 3 (Autumn 1996): 563–82; editorial, "White House Asks to Free Money for Toxic Sites," *New York Times*, 5/18/1998; Jerry Gray, "The Budget Truce: The Details," *New York Times*, 4/26/1996; letter to the editor by Newt Gingrich, "Green, Green G.O.P.," *New York Times*, 5/13/1996.

16. Terry Pristin, "Superfund Item Is Defeated," *New York Times*, 6/20/1997; John H. Cushman, "Deep within Superfund Bill Are Goodies for Montana," *New York Times*, 9/5/1997; editorial, "Environmental Tasks in 1998," *New York Times*, 1/2/1998; editorial, "White House Asks to Free Money for Toxic Sites," *New York Times*, 5/18/1998.

17. Letter from EPA to multiple Missoula agencies and citizen groups, 9/11/1996, reel 932, Milltown Administrative State Site Records.

18. Letter from Larry D. Thompson (MPC) to Bob Fox (EPA), 9/25/1996; letter from EPA to ARCO, 10/9/1996; letter from Meg Nelson (CFC) to Robert Fox and John Wardell (EPA), 11/14/1996; letter from Missoula Board of County Commissioners to Robert Fox (EPA), 11/21/1996; EPA, "Monthly Progress Report, Milltown Reservoir Sediments Superfund Site, January 1997," all on reel 991, Milltown Administrative State Site Records.

19. "4U," special section, "Water Works," *Missoulian*, 5/27/1996.

20. Letter from C. Richard Clough (FWP) to John Wardell (EPA), 5/2/1996, reel 932, Milltown Administrative State Site Records.

21. Dennis Workman (FWP), "A Review of Trout Populations in the Clark Fork River, Warm Springs to Superior, Montana," 4/1985, reel 144; MPC, "Milltown Fisheries Protection, Mitigation, and Enhancement Plan," 10/29/1992, reel 144, Milltown Administrative State Site Records.

22. Clark Fork Coalition, "Comments of the Clark Fork Coalition to the Federal Energy Regulatory Commission," 5/19/1986, reel 150; letter from FWP to FERC, 2/24/1987, reel 169, Milltown Administrative State Site Records.

23. Associated Press, "Study Planned on Blackfoot Fishery," *Missoulian*, 3/19/1988; Dick Manning, "Biologists Set to Solve Blackfoot Fishing Puzzle," *Missoulian*, 3/24/1988; Sherry Jones, "Frayed Fishery: Researchers Look for Clues to Halt Blackfoot's Decline," *Missoulian*, 5/18/1989.

24. Sherry Devlin, "Let 'Em Go," *Missoulian*, 9/29/1989; Bruce Farling, OH 428–04.

25. Richard Manning, *Last Stand: Logging, Journalism, and the Case for Humility* (Salt Lake City: Peregrine Smith Books, 1991), 56.

26. Daryl Gadbow, "Study Divides Blackfoot's Trouble Spots," *Missoulian*, 5/17/1990.

27. Daryl Gadbow and Jeff Herman, "Bad Press May Have Been a Boon to the Blackfoot," *Missoulian*, 7/11/1991; Daryl Gadbow, "Blackfoot on the Mend," *Missoulian*, 7/18/1991.

28. Greg Tollefson, "Flies Right, but the Wrong River Runs through It," *Missoulian*, 9/17/1992; Sherry Devlin, "Blackfoot to Benefit from 'River' Showing," *Missoulian*, 10/10/1992; Carolynn McLuskey, "'River' Rakes in Funds for Rehabilitation," *Missoulian*, 4/5/1993.

29. MPC, "Milltown Fisheries Protection, Mitigation, and Enhancement Plan," 10/29/1992, reel 144, Milltown Administrative State Site Records.

30. Mick Holien, "MPC Thinks Fish Ladder Would Keep Step with Trout," *Missoulian*, 8/26/1993; Sherry Devlin, "Montana Power to Unveil Fishery Enhancement

Plan," *Missoulian*, 3/23/1993; Sherry Devlin, "Milltown Dam Plan Leaves Trout Trapped, Expert Says," *Missoulian*, 3/25/1993. ARCO, "Comments on the Phase I Milltown Fisheries Protection, Mitigation, and Enhancement Plan," 5/28/1993; letter from Steve Schombel (TU) to Don Sprague (MPC), 10/21/1993; letter from C. B. Pearson (CFC) to Don Sprague (MPC), 10/22/1993; MPC, "Final Report: Fish Passage at Milltown Dam, Montana—A Feasibility Study," 9/1993; ARCO, "Comments on the Phase II Milltown Fisheries Protection, Mitigation, and Enhancement Plan," 10/22/1993, all on reel 684, Milltown Administrative State Site Records.

31. Letter from Don Sprague (MPC) to FERC, 12/28/1993, reel 684, Milltown Administrative State Site Records.

32. Salish–Pend d'Oreille Culture Committee, *Salish People*, 46–51; Farr, "Going to Buffalo," part 2, 26-43.

33. R. J. Behnke, *Monograph of the Native Trouts of the Genus* Salmo *of Western North America* (U.S. Forest Service, U.S. Fish and Wildlife Service, and Bureau of Land Management, 1979); Bob Batz and Raissa Marks, Natural Resources Conservation Service, *Bull Trout:* Salvelinus confluentus, Fish and Wildlife Habitat Management Leaflet 36, 1/2006, http://www.wildlifehc.org/new/wp-content/uploads/2010/10/Bull-Trout.pdf; "Bull Trout: About Bull Trout," U.S. Fish and Wildlife Service, accessed 11/16/14, http://www.fws.gov/pacific/bulltrout/About.html.

34. Letter from Michael T. Pablo (CSKT) to Don Sprague (MPC), 10/23/1993, reel 684, Milltown Administrative State Site Records.

35. Sherry Devlin, "Fish Net Boosts Trout's Chances," *Missoulian*, 3/28/1995.

36. Ginny Merriam, "Bringing Back the Big Blackfoot," *Missoulian*, 11/9/1995. For more on the Blackfoot Challenge, including links to new stories on the group's history and achievements, see http://blackfootchallenge.org/Articles/.

37. Sherry Devlin, "Blackfoot on the Brink," *Missoulian*, 4/18/1996; "Governor: Planned Gold Mine May Be Threat to Bull Trout," *Missoulian*, 6/4/1996; "Trout Reports Expected This Week," *Missoulian*, 6/4/1996; Karen Knudsen and Geoff Smith, "You've Heard about It, Now Take Your Own Look," *Missoulian*, 7/21/1996; Paul Larmer, "Bull Trout History," *High Country News*, 5/15/1995.

38. Erin P. Billings, "Mine Study: Fish Populations Depend on Habitat Variables," *Missoulian*, 3/6/1997.

39. Associated Press, "Blackfoot, Missouri Named as Threatened," *Missoulian*, 4/17/1997; Manning, *One Round River*, 182.

40. Daryl Gadbow, "Supporters Push for Bull Trout Listing," *Missoulian*, 7/11/1997; Petersen, *Acting for Endangered Species*.

41. Sherry Devlin, "FWP Biologist Has Been Hardened Thorn in the Side of the Seven-Up Pete Joint Venture," *Missoulian*, 8/21/1997; Daryl Gadbow, "New Bull Trout Spawning Grounds Show Value of Tributaries," *Missoulian*, 8/21/1997; Sherry Devlin, "Mine Puts Blackfoot on Group's Danger List," *Missoulian*, 4/6/1998; Larmer, "Bull Trout History."

42. David Schmetterling, OH 428–06.

43. Daryl Gadbow, "Against the Wall," *Missoulian*, 9/17/1998; Farling, OH 428–04.

44. Schmetterling, OH 428–06. Also see Sherry Devlin, "Trout by Trout: Biologists Carrying Cutthroats up and over Milltown Dam to Help Embattled Fish Complete Their Spawning Cycle," *Missoulian*, 5/27/1999.

45. Sherry Devlin, "A River Run Ragged," *Missoulian*, 3/14/1999; Sherry Devlin, "Protect Blackfoot River but Keep It Open, Agency Told," *Missoulian*, 3/17/1999; Land Lindbergh, guest column, "Restrictions on Recreation May Be Wisest Course to Save the Blackfoot," *Missoulian*, 4/6/1999.

Chapter 6

1. Sherry Devlin, "Milltown Messages Removed from FERC Site," *Missoulian*, 12/24/02; Nielsen, OH 428–08.

2. Nielsen, OH 428–08; Sherry Devlin, "Milltown Messages Removed from FERC Site," *Missoulian*, 12/24/2002; Sherry Devlin, "Feds Issue Milltown Apology," *Missoulian*, 1/10/2003.

3. Tracy Stone-Manning, OH 428–03.

4. Ibid.

5. Sean Benton, OH 428–05.

6. "Montanans Weigh Options on a Toxic Legacy," *New York Times*, 5/7/2001.

7. Benton, OH 428–05. All advertising material created by Partners Creative; copies in possession of author. Sherry Devlin, "Toxics, Trouble Mounting behind Dam," *Missoulian*, 7/11/1997.

8. Russ Forba, personal communication, 6/18/2011, notes in author's possession. Letter from Phyllis Flack (ARCO) to EPA, 11/20/1998, reel 1147; letter from Russ Forba (EPA) to Phyllis Flack (ARCO), 12/16/1998, reel 1147, Milltown Administrative State Site Records; Sherry Devlin, "Milltown Dam a Liability for MPC Sale," *Missoulian*, 4/6/1998; Sherry Devlin, "Engineer Confirms Leaks in Milltown Dam," *Missoulian*, 2/13/2001; Sherry Devlin, "Milltown Dam Remains MPC's Liability," *Missoulian*, 11/3/1998.

9. John Stucke, "BP Amoco to Purchase Arco for $25 billion," *Missoulian*, 3/30/1999.

10. John Stucke, "Judge Oks Arco Deal," *Missoulian*, 4/20/1999.

11. John Stucke, "BP Buyout Could Net Arco Brass Big Bucks," *Missoulian*, 7/20/1999.

12. Prepared for ARCO by Schafer and Associates, "Milltown Reservoir Sediments Site, Draft Focused Feasibility Study," reel 1194, Milltown Administrative State Site Records.

13. Letter from Russell Forba (EPA) to Robin Bullock (ARCO), 11/8/1999, reel 1194, Milltown Administrative State Site Records.

14. Sherry Devlin, "Groups to Oppose MPC Dam Relicensing," *Missoulian*, 2/23/2000.

15. Montana Power Company, "MPC and ARCO Sign Milltown Dam Stewardship Agreement," 5/2/2000, reel 1372, Milltown Administrative State Site Records.

16. Opinion, "Shortchanging Superfund," *New York Times*, 7/6/2002; Missoula Board of County Commissioners, "Case for Removal of Milltown Dam," 2/1/2000, reel 1282; letter from Missoula City-County Board of Health to FERC, 5/17/2001, reel 1371, Milltown Administrative State Site Records; Nielsen, OH 428–08.

17. All advertising material created by Partners Creative; copies in possession of author. Clark Fork Coalition, "Cleanup Myths," *Currents*, Spring 2001, reel 1372, Milltown Administrative State Site Records.

18. All advertising material created by Partners Creative; copies in possession of author.

19. U.S. EPA and Montana Department of Environmental Quality, "Superfund Clean-Up Proposal: Milltown Reservoir Sediments Operable Unit of the Milltown Reservoir/Clark Fork River Superfund Site" (Helena, Mont.: U.S. Environmental Protection Agency, Region 8, Montana Office, 4/2003), 2; Sherry Devlin, "EPA Offices Flooded with Public Comment on Milltown Dam," *Missoulian*, 10/30/2002.

20. E-mail from John Adams to Wardell (EPA), 7/12/2001; e-mail from Catherine Price-Alick and Claude Alick to EPA, 8/11/2001; e-mail from Lee Torgrimson to Judy Martz, EPA, *Missoula Independent*, *Missoulian*, *Montana Standard* (Butte, Mont.), and *Deer Lodge (Mont.) Silver State Post*, 9/9/2001; e-mail from Dan J. Durham to Martz and EPA, 9/15/2001, all on reel 1472, Milltown Administrative State Site Records.

21. E-mail from Robert Logan to John Wardell, 8/18/2001, reel 1472; e-mail from Brett Hackman to Wardell, 4/29/2002, reel 1472, Milltown Administrative State Site Records.

22. Sherry Devlin, "MPC, Arco Want Cleanup, Not Removal," *Missoulian*, 5/3/2000; Sherry Devlin, "MPC: Buy It All, or Don't Bid," *Missoulian*, 5/6/2000; Sherry Devlin, "Engineer Confirms Leaks in Milltown Dam," *Missoulian*, 2/13/2001; Sherry Devlin, "State Angry over Leak Disclosure" *Missoulian*, 2/17/2001; Sherry Devlin, "Official: Cracks Not a Risk" *Missoulian*, 3/1/2001; Sherry Devlin, "Divers Map Milltown Dam Cracks in Detail," *Missoulian*, 3/2/2001; "Montanans Weigh Options on a Toxic Legacy," *New York Times*, 5/7/2001. Letter from Sandra Stash (ARCO) to Missoula County Board and City Council, 6/11/2001, reel 1372, Milltown Administrative State Site Records.

23. Sherry Devlin, "FERC Won't Reconsider Relicensing Milltown Dam," *Missoulian*, 9/27/2000; Sherry Devlin, "Group Wants Access to Milltown Pact," *Missoulian*, 1/11/2002; Sherry Devlin, "MPC Releases Secret Agreement on Dam," *Missoulian*, 1/26/2002. Letter from ARCO to EPA, 11/16/2000, reel 1331; ARCO Environmental Remediation L.L.C. (prepared by EMC²), "Milltown Reservoir Sediments Site Draft Focused Feasibility Study," 11/15/2000, reel 1282; EMC², "Remedial Alternatives Retained for Further Evaluation in Focused Feasibility Study," 7/14/2000, reel 1282; letter from Michael Manion (MPC) to Russ Forba (EPA), 1/3/2001, reel 1371, Milltown Administrative State Site Records.

24. Letter from Matt Clifford (CFC) to Russ Forba (EPA), 1/2/2001, reel 1372; letter from Peter Nielsen (MCCHD) to Forba, 1/2/2001, reel 1372; Dennis Workman,

"Partial Comments by CFRTAC on the Milltown Reservoir Sediments Site Draft Focused Feasibility Study," 1/2/2001, reel 1371; letter from Don Skaar, Ladd Knotek, David Schmetterling (FWP) to Forba, 1/4/2001, reel 1372, Milltown Administrative State Site Records.

25. Letter from Johnnie Moore and William Woessner (UM) to Arvid Hiller (Mountain Water Company), 12/21/2000; letter from Arvid Hiller (MWC) to Russ Forba (EPA), 12/27/2000; letter from USFWS to Forba, 1/8/2001; U.S. Army Corps of Engineers, "Comments on Milltown Dam Superfund Site-Focused Feasibility Study," 2/21/2001; EPA, "Comments on the Milltown Reservoir Sediments Site Revised Final FFS," 4/27/2001; letter from Paul Schroeder (U.S. Army Engineer Research and Development Center) to Forba, 8/10/2001, all on reel 1371, Milltown Administrative State Site Records; Sherry Devlin, "Caution Urged in Cleanup," *Missoulian*, 8/30/2001; Sherry Devlin, "Corps' Milltown Dredging Report Outlined," *Missoulian*, 6/20/2001; Sherry Devlin, "Milltown Dredging Deemed 'Fairly Clean,'" *Missoulian*, 8/14/2001.

26. Letter from Russ Forba (EPA) to Robin Bullock (ARCO), 7/3/2001, reel 1372; U.S. Army Corps of Engineers, "Technical Report: Evaluation of Dam Removal Costs," 10/31/2000, reel 1331, Milltown Administrative State Site Records; U.S. Army Corps of Engineers, "Overview of Dam Decommissioning," accessed October 12, 2011, https://web.archive.org/web/20090110004834/http://www.crrel.usace.army.mil/sid/Dam_decom/overview.htm.

27. E-mail from Stuart Goldberg to Russ Forba (EPA), 6/30/2000, reel 1282; letter from David Ausband to Forba, 1/27/01, reel 1331; e-mail from Patia Stephens to John Wardell (EPA), 7/1/2001, reel 1472; e-mail from Christopher C. Ballentine to EPA, 7/9/2001, reel 1472; e-mail from Richie Moore to EPA, 6/16/2001, reel 1472, Milltown Administrative State Site Records; John Stromnes, "Sediment Won't Harm Residents," *Missoulian*, 9/15/2001.

28. Letter from Mark Everingham to Forba, 8/4/2000, reel 1282; letter from Rosanne E. Davis to Forba, 12/7/2000, reel 1331; e-mail from Elise Crull to EPA, 7/5/2001, reel 1472, Milltown Administrative State Site Records.

29. E-mail from Doug Anderson to EPA and Martz, 10/9/2001, reel 1472, Milltown Administrative State Site Records.

30. Benton, OH 428–05.

31. All advertising material created by Partners Creative; copies in possession of author.

32. John H. Cushman, Jr., "E.P.A. Proposes New Rule to Lower Arsenic in Tap Water," *New York Times*, 5/25/2000.

33. Letter from L. F. Schombel to Russ Forba (EPA), 6/6/2000, reel 1282, Milltown Administrative State Site Records.

34. Douglas Jehl, "E.P.A. to Abandon New Arsenic Limits for Water Supply," *New York Times*, 3/21/2001; Chuck Fox, "Arsenic and Old Laws," *New York Times*, 3/22/2001; Robin Toner, "Environmental Reversals Leave Moderate Republicans Hoping for Greener Times," *New York Times*, 4/4/2001; Douglas Jehl, "E.P.A. Delays

Its Decision on Arsenic," *New York Times*, 4/19/2001; Katherine Q. Seelye, "Arsenic Standard for Water Is Too Lax," *New York Times*, 9/11/2001; John K. Wiley, "No Breaching of Dams, Bush Vows," *Seattle Times*, 2/26/2000.

35. Sherry Devlin, "EPA to Crack Down on Arsenic Levels," *Missoulian*, 2/20/2001; Katherine Q. Seelye, "E.P.A. to Adopt Clinton Arsenic Standard," *New York Times*, 11/1/2001.

36. All advertising material created by Partners Creative; copies in possession of author.

37. Board of Missoula county commissioners, "Two Rivers Restoration and Development Project," n.d., reel 1331, Milltown Administrative State Site Records.

38. Sherry Devlin, "Restoring Nature's Flow," *Missoulian*, 2/22/2001; Sherry Devlin, "County Unveils River Restoration Proposal," *Missoulian*, 2/22/2001; Sherry Devlin, "Two Rivers Talk," *Missoulian*, 3/15/2001. Letter from Michael Enk (Montana chapter of American Fisheries Society) to Missoula county commissioners, 3/6/2001, reel 1331; letter from Triel Culver (American Whitewater Association) to John Wardell (EPA), 6/6/2002, reel 1524; letter from John Anderson (Missoula Whitewater Association) to Wardell, 7/12/2001, reel 1524; letter from Dan Gavere (Wave Sport Kayaks) to Wardell, 3/6/2001, reel 1331; letter from Brian J. Daly to Wardell, 3/6/2001, reel 1331, Milltown Administrative State Site Records.

39. Benton, OH 428–05.

40. Stone-Manning, OH 428–03.

41. Ibid.; Sherry Devlin, "Fleet Floats in Support of Dam's Removal," *Missoulian*, 7/14/2002; Sherry Devlin, "Debating the Dam," *Missoulian*, 7/14/2002.

42. Diana Hammer, OH 428–02; Sherry Devlin, "Corps Will Assess Dam Removal," *Missoulian*, 4/4/2000; letter from Hammer to EPA, 4/10/2003, reel 1709, Milltown Administrative State Site Records.

43. EPA Region 8, "Milltown Reservoir Sediments Superfund Site: Combined Feasibility Study," 4/2002, reel 1472, Milltown Administrative State Site Records.

44. Hilary Sigman, "The Pace of Progress at Superfund Sites: Policy Goals and Interest Group Influence," *Journal of Law and Economics* 44 (April 2001): 315.

45. Kurt D. Cline, "Influences on Intergovernmental Implementation: The States and the Superfund," *State Politics and Policy Quarterly* 3, no. 1 (Spring 2003): 66.

46. William R. Jordan III, and George M. Lubick, *Making Nature Whole: A History of Ecological Restoration* (Washington, D.C.: Island Press, 2011), 3, 135, 147, 173.

47. Rebecca Lave, *Fields and Streams: Stream Restoration, Neoliberalism, and the Future of Environmental Science* (Athens: University of Georgia Press, 2012), 1, 18. Also, "Water: Adopt Your Watershed," U.S. Environmental Protection Agency, accessed 4/29/2014, http://water.epa.gov/action/adopt/index.cfm.

48. Benton, OH 428–05; John Stucke, "Judge Oks Arco Deal," *Missoulian*, 4/20/1999.

49. Sherry Devlin, "Company: Sediments Must Stay," *Missoulian*, 11/29/2000; Sherry Devlin, "Officials, Conservationists Say Time Is Now for Reservoir Cleanup," *Missoulian*, 2/6/2001.

50. Letter from Stone-Manning to EPA, 2/13/2001, reel 1331; e-mail from Matt Desch to Forba, 2/27/2001, reel 1331, Milltown Administrative State Site Records; Sherry Devlin, "Hearing on Dam Planned," *Missoulian*, 3/14/2001; Sherry Devlin, "Sediment Must Go, Board Says," *Missoulian*, 3/29/2001; editorial, "Do the Milltown Reservoir Cleanup Right," *Missoulian*, 3/30/2001.

51. Farling, OH 428–04; Nielsen, OH 428–08; Stone-Manning, OH 428–03.

52. Stone-Manning, OH 428–03.

53. Farling, OH 428–04.

54. Sherry Devlin, "Officials Say Agencies Need to Take Actions," *Missoulian*, 5/31/2000; Barbara Evans, Bill Carey, and Jean Curtis, "Time for Super Cleanup," *Missoulian*, 4/1/2001; Board of County Commissioners, "Resolution 2001–030," 3/28/2001, reel 1371, Milltown Administrative State Site Records.

55. Palmer, *Endangered Rivers*, 304. On the partisan politicization of environmentalism, see Rothman, *Greening of a Nation*; and Wellock, *Preserving the Nation*.

56. E-mail from Steven E. Bristol to EPA, 6/8/2001, reel 1472; e-mail from Jeff Rolston-Clemmer to Wardell, 7/9/2001, reel 1472, Milltown Administrative State Site Records.

57. Letter from Jerry Driscoll, Montana State AFL-CIO, to Judy Martz, 6/11/2001; e-mail from Brad Robinson to Martz and EPA, 7/20/2001; e-mail from Kirsten Renander to Martz and EPA, 9/23/2001, all on reel 1472, Milltown Administrative State Site Records.

58. Sherry Devlin, "Debating the Dam," *Missoulian*, 7/14/2002; letters from Judy Martz to Gary W. Hawk and Dean Ritz, 8/15/2001, reel 1372, Milltown Administrative State Site Records.

59. U.S. EPA, Region 8, Montana Office, "EPA Releases Final Combined Feasibility Study of the Milltown Reservoir," 11/5/2001, reel 1709, Milltown Administrative State Site Records; Sherry Devlin, "Arco Recommends Leaving Milltown Sediment," *Missoulian*, 11/6/2001.

60. Sherry Devlin, "Arco Recommends Leaving Milltown Sediment," *Missoulian*, 11/6/2001; U.S. EPA, Region 8, Montana Office, information postcard including web page fact sheet address and contact information, reel 1709, Milltown Administrative State Site Records.

61. Letter from Jerry Tavegia (Montana Department of Commerce) to Gary Matson (BDG), 9/13/1996, reel 1524, Milltown Administrative State Site Records.

62. John Stucke, "Group Wants Milltown Dam Left Alone," *Missoulian*, 2/7/1999; letter from Bruce Hall (BDG) to gubernatorial candidates, 4/27/2000, reel 1282; letter from Hall to Forba (EPA), 12/11/2000, reel 1331; letter from Board of Directors, Bonner Development Group to Forba, 12/21/2000, reel 1372, Milltown Administrative State Site Records.

63. E-mail from Daniel R. Monroe to EPA, *Missoulian*, and Montana state governor, 6/27/2001, reel 1472; letter from Elden Inabnit to *Missoulian*, 9/6/2001, reel 1524; letter from Don Wibracht to Wardell, 8/22/2002, reel 1472, Milltown Administrative State Site Records.

64. E-mail from Leroy Zent to Judy Martz, 7/19/2001, reel 1472; letter from Alan T. Hoyt to Wardell, 7/11/2001, reel 1524; e-mail from Gary Bray to EPA, 11/16/2001, reel 1472, Milltown Administrative State Site Records.

65. Letter from Bruce Hall (BDG) to Wardell (EPA), 10/29/2001, reel 1418; letter from Hall to Diana Hammer (EPA), 11/5/2001, reel 1524, Milltown Administrative State Site Records.

66. Letter from Tracy Stone-Manning to Wardell, 12/21/2001, reel 1524; letter from Nielsen (MCCHD) to Forba, 1/3/2002, reel 1417; letter from Missoula Board of County Commissioners to Wardell, 1/3/2002, reel 1417, Milltown Administrative State Site Records; Sherry Devlin, "County Tells EPA That Long-Term Cleanup of Milltown Is Most Important," *Missoulian*, 1/5/2002; Sherry Devlin, "Ambrose Contributes to Milltown Fund," *Missoulian*, 1/16/2002.

67. Sherry Devlin, "Fund Astray: 'Mother of Superfund' Says Program Had Good Start, but Is Now a 'Mess,'" *Missoulian*, 3/12/2002.

68. Sherry Devlin, "'Unique Opportunity,'" *Missoulian*, 1/23/2002.

69. Letter from Judy Martz to Ted Antonioli (Rocky Mountain chapter, Montanans for Multiple Use), 3/22/2001, reel 1418, Milltown Administrative State Site Records; Sherry Devlin, "Removal Focus of EPA Talks," *Missoulian*, 3/28/2002.

70. Sherry Devlin, "Tribes Weigh In, Say Dam Should Be Dismantled," *Missoulian*, 3/28/2002; letter from Robert T. Anderson (Division of Indian Affairs, U.S. Department of the Interior) to Michael T. Pablo (CSKT), 12/4/1995, reel 1418; letter from Fred Matt to Wardell, 3/15/2002, reel 1524; CSKT, "U.S.A. before the FERC: Motion to Intervene of the Confederated Salish and Kootenai Tribes," 3/18//2002, reel 1418; letter from Matt to Martz, 6/28/2002, reel 1524, Milltown Administrative State Site Records.

71. "42 U.S. Code § 9626—Indian Tribes," Cornell University Law School, accessed 11/12/2014, http://www.law.cornell.edu/uscode/text/42/9626.

72. EPA news release, "EPA Announces $1.3 Million in Superfund Funding to Return Contaminated Sites to Productive Use," 7/18/2002, reel 1711, Milltown Administrative State Site Records.

73. Office of Congressional and Intergovernmental Relations, "Milltown, Montana," n.d., reel 1711; Missoula County press release, "$40,000 Awarded for Milltown-Bonner Area Superfund Redevelopment," 7/26/2002; reel 1711, Milltown Administrative State Site Records; Sherry Devlin, "Federal Funds Allocated for Milltown Planning," *Missoulian*, 7/31/2002; "Superfund: Superfund Redevelopment Initiative," U.S. Environmental Protection Agency, http://www.epa.gov/superfund/programs/recycle/.

74. Letter from Hall (BDG) to Wardell and Missoula Board of County Commissioners, 8/20/2002, reel 1524; EPA, "Superfund Redevelopment Pilot for the Milltown Reservoir Sediments Superfund Site," 8/26/2002, reel 1711; letter from Hall to Wardell, Martz, et al., 8/29/2002, reel 1524; EPA, "Milltown Superfund Site Redevelopment and Restoration Initiative," 9/2002, reel 1524; letter from Missoula Board of County Commissioners to Diana Hammer, 10/22/2001, reel 1711; James Ariail, Catherine Allen, and Linda Manning, "Proposed Project Approach Facilitating the

Formation of a Working Group to Advise EPA Region 8 on the Future Use of the Milltown Reservoir Sediment Site," 10/28/2002, reel 1711, Milltown Administrative State Site Records.

75. Sherry Devlin, "Milltown Dam Drawdown Begins Monday," *Missoulian*, 8/2/2002; Sherry Devlin, "Superfund Official Inspects Reservoir to Get Better Idea of Cleanup Costs," *Missoulian*, 8/23/2002.

76. EPA, "Environmental News: EPA Proposes Cleanup Plan for the Clark Fork River," 8/15/2002, reel 1473, Milltown Administrative State Site Records; Matt Gouras, "EPA Unveils Clark Fork Cleanup Plan," *Missoulian*, 8/16/2002; Sherry Devlin, "Healing at Hand," *Missoulian*, 9/12/2002.

77. E-mail from Diana Hammer (EPA) to EPA, 9/10/2002; letter from Mary Erickson to EPA, 10/8/2002; letter from Gary Matson to EPA, 10/12/2002, all on reel 1524, Milltown Administrative State Site Records; Sherry Devlin, "New Group to Back Milltown Dam Removal," *Missoulian*, 10/24/2002; Sherry Devlin, "EPA Offices Flooded with Public Comments on Milltown Dam," *Missoulian*, 10/30/2002; Greg Matson, "Guest Column: Restoring the River Would Revitalize a Community, Provide Permanent Remedy," *Missoulian*, 11/6/2002.

78. Letter from Stone-Manning to Wardell et al., 10/30/2002, reel 1525; EPA, "Press Advisory: Milltown Reservoir Clean Up Decision Schedule Announced," 12/23/2002, reel 1524, Milltown Administrative State Site Records.

79. Stone-Manning, OH 428–03.

80. Ibid.; letter from Martz to Bruce Hall (BDG), 12/24/2002, reel 1524, Milltown Administrative State Site Records; Sherry Devlin, "State of the State Address: Martz: Remove Milltown Dam," *Missoulian*, 1/22/2003; Sherry Devlin, "EPA Wants Milltown Dam Removed," *Missoulian*, 1/23/2003; Blaine Harden, "Once Again, a River Will Run through It," *Washington Post*, 4/16/2003.

81. Stone-Manning, OH 428–03.

82. Sherry Devlin, "Baucus in Bonner to Talk about Milltown," *Missoulian*, 2/19/2003; Sherry Devlin, "Baucus Offers Help," *Missoulian*, 2/20/2003; Sherry Devlin, "'Right Thing to Do,'" *Missoulian*, 2/28/2003; letter from Martz to Missoula county commissioners, 3/5/2003, reel 1524, Milltown Administrative State Site Records.

83. Farling, OH 428–04; Stone-Manning, OH 428–03.

84. Sherry Devlin, "Dam's Days Numbered," *Missoulian*, 4/16/2003; e-mail from Diana Hammer to EPA, 4/10/2003, reel 1709; U.S. EPA, Region 8, news release, "EPA and DEQ Release Proposed Cleanup Plan for Milltown Reservoir Sediments Superfund Site," 4/15/2003, reel 1524, Milltown Administrative State Site Records.

85. Sherry Devlin, "Public Urged to Comment on Details of River's Future," *Missoulian*, 4/16/2003; Hammer, OH 428–02. For a statistical study of the EPA's preference for permanent cleanup plans, see Gupta, Van Houtven, and Cropper, "Paying for Permanence."

86. Hammer, OH 428–02.

87. "Superfund: Natural Resource Damages; A Primer," U.S. Environmental Protection Agency, accessed 5/15/2014, http://www.epa.gov/superfund/programs/nrd

/primer.htm; also see U.S. EPA, *Integrating Water and Waste Program to Restore Watersheds: A Guide for Federal and State Project Managers* (Washington D.C.; Office of Water and Office of Solid Waste and Emergency Response, U.S. Environmental Protection Agency, 8/2007), 47–48, 176–78.

88. U.S. EPA, *Integrating Water and Waste Program*, i, 176–78.

89. "Superfund: Superfund and Green Remediation," U.S. Environmental Protection Agency, 9/2010, accessed 5/15/2014, http://www.epa.gov/superfund/green remediation/sf-gr-strategy.pdf, p. 10.

90. See "The Three R's of the Milltown Reservoir Superfund Project," Clark Fork River Technical Assistance Committee, https://web.archive.org/web/20140219202853/http://www.cfrtac.org/061009b.html. The EPA developed websites devoted to its restoration and redevelopment processes and accomplishments: "Natural Resource Damages: A Primer," http://www.epa.gov/superfund/programs/nrd/primer.htm; and "Superfund Redevelopment Initiative," http://www.epa.gov/superfund/programs/recycle/index.html.

91. Editorial, "Milltown Decision Will Prove Momentous," *Missoulian*, 4/22/2003.

Chapter 7

Epigraph: "Bruce Babbitt," American Academy for Park and Recreation Administration, accessed 3/16/2012, http://www.aapra.org/pugsley-bios/bruce-babbitt.

1. Christine Brick (CFC), "Clark Fork River Restoration," lecture at the University of Montana, Missoula, 3/27/2012.

2. Dan Flores, "The West That Was, and the West That Can Be: Western Restoration and the Twenty-First Century," in *Natural West*; Elizabeth Grossman, *Watershed: The Undamming of America* (New York: Counterpoint, 2002), 3.

3. "Integrating the '3 Rs': Remediation, Restoration, and Redevelopment," U.S. Environmental Protection Agency, http://www.epa.gov/superfund/programs/recycle/pdf/milltown-casestudy.pdf.

4. Joe Truini, "Big Dam Cleanup in Big Sky Country," *Waste News*, 4/28/2003.

5. John Cramer, "Into the Breach: Clark Fork, Blackfoot Rivers Punch through Milltown Dam" *Missoulian*, 3/29/2008; Sherry Devlin, "Dignitaries Hail Return of Unhindered Flows," *Missoulian*, 3/29/2008; personal observations by the author; Diana Hammer interview, OH 428–02.

6. See Ellen Stroud, "Troubled Waters in Ecotopia: Environmental Racism in Portland, Oregon," in *American Environmental History*, ed. Louis S. Warren (Malden, Mass.: Blackwell Publishing, 2003); Mark Dowie, *Losing Ground: American Environmentalism at the Close of the Twentieth Century* (Cambridge, Mass.: MIT Press, 1995); Deeohn Ferris and David Hahn-Baker, "Environmentalists and Environmental Justice Policy," in *Environmental Justice: Issues, Policies, and Solutions*, ed. Bunyan Bryant (Washington, D.C.: Island Press, 1995).

7. EMC[2], "Disposal Location Sensitivity Analysis," 7/25/2000, reel 1282, Milltown Administrative State Site Records.

8. Letter from Peter Nielsen (MCCHD) to Russ Forba (EPA), 8/21/2000, reel 1282, Milltown Administrative State Site Records.

9. Farling, OH 428–04.

10. Edwin Dobbs, "Pennies from Hell: In Montana, the Bill for America's Copper Comes Due," *Harper's Magazine*, October 1996, 45. Letter from John Wardell to Dan Watts, 4/9/2002; letter from Watts to Wardell, 4/16/2002; letter from Watts to Wardell, 6/10/2002, all on reel 1472, Milltown Administrative State Site Records.

11. Sherry Devlin, "Disposal Debate: Forum Addresses Possible Sites for Milltown Sediments," *Missoulian*, 5/1/2002; Sherry Devlin, "Dam's Days Numbered," *Missoulian*, 4/16/2003; U.S. EPA, "EPA and DEQ Release Proposed Cleanup Plan for Milltown Reservoir Sediments Superfund Site," 4/15/2003, reel 1524, Milltown Administrative State Site Records.

12. E-mail from Diana Hammer to EPA, 6/17/2003, reel 1709; U.S. EPA, "Public Comment Period Extended," 6/17/2003, reel 1709, Milltown Administrative State Site Records.

13. Sherry Devlin, "Deadline Extended to Comment on Milltown Cleanup," *Missoulian*, 6/21/2003; Sherry Devlin, "Public Urged to Comment on Details of River's Future," *Missoulian*, 4/16/2003; e-mail from Brooks and Jackie Sanford to EPA, 4/16/2003, reel 1709, Milltown Administrative State Site Records.

14. Jim Robbins, "In a Town Called Opportunity, Distress Over a Dump," *New York Times*, 8/24/2005.

15. Martin-Lake & Associates, Inc., The Court Reporters, "Superfund Program Clean-Up Proposal Milltown Reservoir/Clark Fork River, Transcript of Proceedings," 5/7–5/8/2003, reel 1709, Milltown Administrative State Site Records.

16. Brad Tyer, *Opportunity, Montana: Big Copper, Bad Water, and the Burial of an American Landscape* (Boston: Beacon Press Books, 2013), 118–19.

17. Letter from Max Baucus to Russ Forba, 5/5/2003; letter from Sherm Anderson to Russ Forba, 6/8/2003; e-mail and letter from G. Bruce Morris (Carpenter's Local Union No. 28) to Forba, 6/18/2003; letter from Montana Community-Labor Alliance/Jobs with Justice to Forba, 7/15/2003, all on reel 1709, Milltown Administrative State Site Records; Ashley Fingarson, "Cloud with a Copper Lining: From Superfund Site to Golf Course," Property and Environment Research Center, http://www.perc.org/articles/article220.php; "Anaconda Co. Smelter," U.S. Environmental Protection Agency, accessed 7/9/2014, http://www2.epa.gov/region8/anaconda-co-smelter.

18. Letter from Gary Carnefix (Carnefix Ecological Consulting) to Russ Forba, n.d. (but references the comment period), reel 1709; letter from Nielsen to Forba, 7/21/2003, reel 1709, Milltown Administrative State Site Records.

19. Sherry Devlin, "Ready to Go: Envirocon Answers Questions on Cleanup Plans," *Missoulian*, 7/23/03.

20. Roberta Forsell Stauffer, "Opportunity Residents Wary of Milltown Waste," *Montana Standard*, 7/29/2003, reel 1710; letter from Charlene Hagan to Forba, 6/18/2004, reel 1710, Milltown Administrative State Site Records; Farling, OH 428–04.

21. Letter from Martz to Wardell, 9/22/2003, reel 1572; U.S. EPA, "Superfund Program Revised Proposed Cleanup Plan: Milltown Reservoir Sediments Operable Unit," 5/2004, reel 1709; EPA press release, "EPA and DEQ Announce Record of Decision for Cleaning Up the Clark Fork River in Montana," 5/4/2004, reel 1670, Milltown Administrative State Site Records; Sherry Devlin, "EPA Unveils Clark Fork Cleanup Plan," *Missoulian*, 5/5/2004; Sherry Devlin, "Milltown Revisions Unveiled," *Missoulian*, 5/18/2004.

22. Letter from Jim Flynn to Forba, 6/8/2004; letter from Kuipers and Associates, LLC, to CFRTAC, 6/16/2004; letter from Montana Community-Labor Alliance/Jobs with Justice, CFC, et al. to Forba, 6/21/2004; letter from Stone-Manning to Wardell, 6/21/2004; letter from Dan Cox to Forba, 6/18/2004; letter from Charlene Hagan to Forba, 6/18/2004, all on reel 1710, Milltown Administrative State Site Records.

23. Perry Backus, "Dirty Subject," *Missoulian*, 10/1/2006.

24. Sherry Devlin, "Revised Cleanup Plan Gets Initial Public Review," *Missoulian*, 6/10/2004; Perry Backus, "Shipping Sediment," *Missoulian*, 10/3/2007.

25. George Plaven, "Concept in Place to Treat Milltown Sediments at Opportunity," *Montana Standard*, 2/1/2012; Dennis R. Neuman, KC Harvey Environmental LLC, "Remediation of the Milltown Dam Sediments in Montana," presentation at 2014 annual conference, Mine Design, Operations and Closure, http://www.mtech.edu/mwtp/conference/2014_presentations/dennis-neuman.pdf.

26. Letter from NorthWestern Corp. to Forba, 7/21/2003, reel 1709, Milltown Administrative State Site Records; Sherry Devlin, "Feds File Objection over Dam Cleanup," *Missoulian*, 11/15/2003; Sherry Devlin, "Groups Hammer Out Milltown Agreement," *Missoulian*, 5/8/2004; Sherry Devlin, "Milltown Pledge Gets OK," *Missoulian*, 6/22/2004; U.S. Department of Justice, "Civil Action No. CV89–039-BU-SEH: Consent Decree for the Milltown Site," U.S. District Court for the District of Montana, Butte Division, 7/12/2005, http://www2.epa.gov/sites/production/files/2013-12/documents/anacondasmelterorder-2013.pdf; Perry Backus, "Bringing Down the Dam: Champagne Popped as Accord Is Finalized," *Missoulian*, 8/3/2005.

27. Sherry Devlin, "Milltown Dam Removal Plan Finalized," *Missoulian*, 12/21/2004; Colin McDonald, "Missoula on Ice," *Missoulian*, 1/12/2005; U.S. EPA, "Superfund Program Revised Proposed Cleanup Plan: Milltown Reservoir Sediments Operable Unit," 5/2004, reel 1709, Milltown Administrative State Site Records.

28. E-mail from Jon Van Arsdale to EPA, 6/20/2003; e-mail from Erin Kalagassy to EPA, 6/20/2003; letter from Robert Stewart (U.S. Department of the Interior) to Forba, 7/1/2003; e-mail from Denise Zembrysi to EPA, 7/8/2003, all on reel 1709, Milltown Administrative State Site Records.

29. Jim Robbins, "Dam and Waste Will Go, Freeing Two Rivers," *New York Times*, 8/4/2005. Also see Jim Robbins, "The Copper Mine Ran through It: Tales of a River's Rescue," *New York Times*, 4/1/2003; Blaine Harden, "Once Again, a River Will Run through It," *Washington Post*, 4/16/2003. On the Edwards Dam see "Edwards Dam and Kennebec River Restoration," Natural Resources Council of Maine, accessed 4/10/2012, http://www.nrcm.org/projects-hot-issues/healthy-waters/edwards-dam-and

-kennebec-restoration/, and "The 10th Anniversary of the Removal of Maine's Edwards Dam," American Rivers, accessed 11/13/2014, http://www.americanrivers.org/initiative/dams/projects/the-10th-anniversary-of-the-removal-of-mainea%C2%89uas-edwards-dam/.

30. Rob Chaney, "Milltown Dam Endgame on Track, Say Project Officials," Missoulian, 10/21/2009.

31. Chris Brick, "Rivers Run Free: As Milltown Dam Removal Nears Completion," Clark Fork Technical Assistance Committee newsletter, Fall 2008; "Defining Restoration for the Clark Fork River at Milltown," Clark Fork Technical Assistance Committee newsletter, 2010, accessed 4/4/2012, https://web.archive.org/web/20140219202853/http://www.cfrtac.org/061009b.html.

32. "Jocko River Restoration Project," Confederated Salish and Kootenai Tribes, http://jockoriver.net/Jindex.lasso; "Jocko River Master Plan," Confederated Salish and Kootenai Tribes, accessed 3/14/2012, http://jockoriver.net/j_master.lasso?page=Master Plan&side=mp&subside=prjdsc&-session=JRR:3F99556B07ab41608CqtI3540390; "Germaine White, Bull Trout's Gift," radio interview, KUFM 89.1, Missoula, http://www.prx.org/pieces/63782-germaine-white-bull-trout-s-gift; Vince Devlin, "Around the Bend: CSKT Works to Restore Watershed Back to Its Natural State," Missoulian, 5/6/2010.

33. "Germaine White, Bull Trout's Gift," radio interview, KUFM 89.1, Missoula, http://www.prx.org/pieces/63782-germaine-white-bull-trout-s-gift; Confederated Salish and Kootenai Tribes, Bull Trout's Gift: A Salish Story about the Value of Reciprocity (Lincoln: University of Nebraska Press, 2011); Jenna Cederberg, "Words Alive: Salish Language Dictionary Expands from 186 Pages to 816," Missoulian, 1/10/2011.

34. CSKT, Bull Trout's Gift, 29. For other literature on how Native American stories reflect ideas about environmental impacts or management, see Keith Basso, Wisdom Sits in Places: Landscape and Language among the Western Apache (Albuquerque: University of New Mexico Press, 1996); Shepard Krech, III, The Ecological Indian: Myth and History (New York: W. W. Norton, 1999); Richard White, Roots of Dependency: Subsistence, Environment, and Social Change among the Choctaws, Pawnees, and Navajos (Lincoln: University of Nebraska Press, 1983); and Donald Fixico, "Ethics and Responsibilities in Writing American Indian History," American Indian Quarterly 20 (Winter 1996).

35. Perry Backus, "Let the Water Flow," Missoulian, 10/6/2005; Perry Backus, "'Practice Run,'" Missoulian, 11/3/2005; Perry Backus, "Blackfoot to Flow Free," Missoulian, 10/20/2005; Perry Backus, "Bonner Dam Gone," Missoulian, 11/22/2005; Perry Backus, "Historical Log," Missoulian, 11/27/2005; Perry Backus, "Fishing for Wood," Missoulian, 10/7/2006; Rob Chaney, "Water Logged: Grant-Funded Effort Pulls Up Century-Old Timbers Plus Piers for Safety on the Recreation Corridor," Missoulian, 10/13/2010.

36. Perry Backus, "Milltown Dam: 'Dead in the Water,'" Missoulian, 4/8/2006; Perry Backus, "Bill Contains $5 Million for Milltown Dam Renovation," Missoulian, 7/20/2005; Perry Backus, "Fishing for Answers," Missoulian, 5/2/2006; Perry Backus,

"Drawing Down the Water," *Missoulian*, 6/2/2006; Michael Moore, "County Supports Milltown Proposal," *Missoulian*, 3/3/2005; "Redevelopment: Milltown State Park and Trail Projects Progress," Clark Fork Technical Assistance Committee newsletter, Winter 2011.

37. Robert Struckman, "Deep Cuts in Bonner," *Missoulian*, 10/2/2005; Tyler Christensen, "Stimson to Cut 43 Jobs at Bonner Mill," *Missoulian*, 12/15/2006; Tyler Christensen, "Lumber Market in Flux," *Missoulian*, 5/28/2007; Perry Backus, "Stimson: Cutting Back in Bonner," *Missoulian*, 5/27/2007; Kim Briggeman, "Fortune Faded: Working-Class Town Loses Luster of Yesteryear," *Missoulian*, 9/3/2007.

38. Environmental restoration job search, SimplyHired, accessed 3/15/2012, http://www.simplyhired.com/a/jobs/list/q-environmental+restoration; "Mission and Vision," Society for Ecological Restoration, http://www.ser.org/about/mission-vision. A Google search of "environmental restoration company" (with quotes so that the words appear together in any hit) results in more than 73,000 hits. A single website listed more than 1,200 jobs in environmental restoration, mostly as technicians, engineers, project managers, and the like. Perry Backus, "Interior Official Urges Cooperation on Land Restoration," *Missoulian*, 9/26/2006; Phuong Le, "Ag Secretary Looks to Protect Watersheds on USFS Lands," *Missoulian*, 8/15/2009; William Lowry, *Repairing Paradise: The Restoration of Nature in America's National Parks* (Washington, D.C.: Brookings Institution Press, 2009).

39. Benton, OH 428–05; Robert Struckman, "Cooling Off, Cashing In," *Missoulian*, 7/24/2005; Mea Andrews, "Making Waves," *Missoulian*, 1/13/2006; Perry Backus, "Forum to Address Impact of Restoration," *Missoulian*, 5/23/2006; Tyler Christensen, "Making Waves," *Missoulian*, 4/15/2007; Rob Chaney, "Restoration River: Unlocking Potential," *Missoulian*, 9/8/2009; editorial, "Tourism District a Great Promotional Tool," *Missoulian*, 10/16/2009; "Author Speaks on Restoration," Clark Fork Technical Assistance Committee newsletter, June 2006; Norma Nickerson, Christine Oschell, Lee Rademaker, and Robert Dvorak, *Montana's Outfitting Industry: Economic Impact and Industry-Client Analysis* (Missoula: Institute for Tourism and Recreation Research, University of Montana, 3/2007); Al Ellard, Kristin Aldred Cheek, and Norma Nickerson, *Missoula Case Study: Direct Impact of Visitor Spending on a Local Economy* (Missoula: Institute for Tourism and Recreation Research, University of Montana, 4/1999).

40. Thomas Michael Power, *Lost Landscapes and Failed Economies: The Search for a Value of Place* (Washington, D.C.: Island Press, 1996), 160; "State and County Quick Facts," U.S. Census Bureau, http://quickfacts.census.gov/qfd/states/30/3050200.html.

41. Pacific Western Technologies, Ltd., for EPA Region 8, *Water Supply and Milltown Reservoir Sediments Operable Units of the Milltown Reservoir Sediments/Clark Fork River Superfund Site, Missoula County, Montana, First Five-Year Review Report* (Washington, D.C.: Office of Ecosystems Protection and Remediation, U.S. Environmental Protection Agency, 9/2011), 8; U.S. EPA, Region 8, *Clark Fork River Operable Unit of the Milltown Reservoir/Clark Fork River Superfund Site, Record of*

Decision, Part 2: Decision Summary, 4/2004, http://powellcountymt.gov/janda/files/home/1380558283_CFRecofDecisionPt2.pdf, pp. 2–144.

42. Betsy Cohen, "Watershed Moments," *Missoulian*, 9/7/2006; Joe Nickell, "Museums Join Forces to Document Removal," *Missoulian*, 4/22/2006; *Changing Currents, Altered Landscapes* (Missoula: University of Montana Gallery of Visual Arts, autumn 2006). Also personal observation, notes in author's possession.

43. Editorial, "Tiny Trout May Be Harbinger of Rebirth," *Missoulian*, 10/29/2006; Rob Chaney, "Open for Discussion: FWP Wants Input on Public Area Proposals around Former Milltown Dam Site," *Missoulian*, 11/5/2010; Rob Chaney, "Bull Trout: Fish Thriving in Seeley," *Missoulian*, 11/27/2011; Rob Chaney, "Speakers: Unexpected Water to Harbor Good Fish," *Missoulian*, 3/14/2012.

44. Rob Chaney, "Dirt Gone Barren: Milltown Sediment Spread at Opportunity Won't Grow Grass," *Missoulian*, 10/31/2010; Rob Chaney, "Missoula, Butte Both Interested in Restoration Funds," *Missoulian*, 12/5/2010; George Plaven, "Milltown Dam Sediment: Concept for Treating Tailings in Place," *Montana Standard*, 2/2/2012.

45. Hammer, OH 428–02.

46. U.S. EPA, *Integrating the "3 Rs": Remediation, Restoration and Redevelopment, the Milltown Reservoir Sediments Site and Missoula County, Montana* (Denver, Colo.: EPA Region 8 and Helena, Mont.: EPA Montana Operations Office, 4/2011), 1, 10–11.

47. Rob Chaney, "Rush before Runoff," *Missoulian*, 4/26/2011; Rob Chaney, "Channel Change: High Spring Runoff Tests New Direction of Clark Fork River," *Missoulian*, 8/31/2011.

Conclusion

1. For a map of hazardous waste sites in the United States, see "Cleanups in My Community," U.S. Environmental Protection Agency, accessed 4/4/2012, http://iaspub.epa.gov/apex/cimc/f?p=255:63:1603772554611628.

2. Hammer, OH 428–02.

3. Ibid.

4. Farling, OH 428–04; Jenny Price, "Remaking American Environmentalism: On the Banks of the L.A. River," *Environmental History* 13 (July 2008): 536–55.

5. Rob Chaney, "208 Years after Lewis' Travels, Blackfoot an Undaunted Current for Floaters," *Missoulian*, 7/3/2014.

Bibliography

Archival Collections

Charles Herbert McLeod Papers, 1865–1953, MSS 001, Archives and Special Collections, Mansfield Library, University of Montana.

David Brooks's Milltown Oral History Project, 2009–2011, OH 428, Archives and Special Collections, Mansfield Library, University of Montana, Missoula.

Jack L. Demmons / Bonner School Photographs, Archives and Special Collections, Mansfield Library, University of Montana, Missoula, http://www.lib.umt.edu/digital/demmons.

Milltown Administrative State Site Records, U.S. Environmental Protection Agency, Government Documents, Mansfield Library, University of Montana.

Milltown Collection, Missoula City-County Health Department (MCCHD), Missoula, Montana.

Milltown Oral History Collection, OH 419, Archives and Special Collections, Mansfield Library, University of Montana, Missoula

Missoula Light and Water Company Records, 1885–1909, MSS 581, Archives and Special Collections, Mansfield Library, University of Montana.

Montana Power Company (MPC) Predecessor Company Records, 1880–1947, MC 268, Montana Historical Society Archives, Helena, Montana.

Montana Power Company Records, 1892–1967, MSS 240, Archives and Special Collections, Mansfield Library, University of Montana.

Montanans at Work Oral History Project, OH 140, Archives and Special Collections, Mansfield Library, University of Montana, Missoula.

Montanans at Work Oral History Project, OH 266 and OH 392, Montana Historical Society, Helena, Montana.

Books, Articles, and Theses

Adler, Robert W., Jessica C. Landman, and Diane M. Cameron. *The Clean Water Act 20 Years Later*. Washington, D.C.: Island Press, 1993.

Alt, David D., and Donald W. Hyndman. *Roadside Geology of Montana*. Missoula, Mont.: Mountain Press, 1986.

American Social History Project. *Who Built America? Working People and the Nation's Economy, Politics, Culture, and Society*. Vol. 2, *From the Gilded Age to the Present*. New York: Pantheon Books, 1992.

Bailey, Ann K. "Concentration of Heavy Metals in the Sediments of a Hydroelectric Impoundment." Master's thesis, University of Montana, 1976.

Basso, Keith. *Wisdom Sits in Places: Landscape and Language among the Western Apache*. Albuquerque: University of New Mexico Press, 1996.

Behnke, R. J. *Monograph of the Native Trouts of the Genus* Salmo *of Western North America*. U.S. Forest Service, U.S. Fish and Wildlife Service, and Bureau of Land Management, 1979.

Bicentennial Committee Bonner School. *A Grassroots Tribute: The Story of Bonner, Montana*. Missoula, Mont.: Gateway Press, 1976.

Blum, Elizabeth D. *Love Canal Revisited: Race, Class, and Gender in Environmental Activism*. Lawrence: University Press of Kansas, 2008.

Bodnar, John. *The Transplanted: A History of Immigrants in Urban America*. Bloomington: Indiana University Press, 1985.

Brooks, David. *Bobos in Paradise: The New Upper Class and How They Got There*. New York: Simon and Schuster, 2000.

Bryant, Bunyan, ed. *Environmental Justice: Issues, Policies, and Solutions*. Washington, D.C.: Island Press, 1995.

Clark, Ella E. *Indian Legends from the Northern Rockies*. Norman: University of Oklahoma Press, 1966.

Confederated Salish and Kootenai Tribes. *Bull Trout's Gift: A Salish Story about the Value of Reciprocity*. Lincoln: University of Nebraska Press, 2011.

Coon, Shirley. "The Economic Development of Missoula, Montana." Master's thesis, University of Chicago, 1926.

Dobb, Edwin. "Pennies from Hell." *Harper's Magazine*, October 1996.

Dowie, Mark. *Losing Ground: American Environmentalism at the Close of the Twentieth Century*. Cambridge, Mass.: MIT Press, 1995.

Ellard, Al, Kristin Aldred Cheek, and Norma Nickerson. *Missoula Case Study: Direct Impact of Visitor Spending on a Local Economy*. Missoula: Institute for Tourism and Recreation Research, University of Montana, 1999.

Environment and Natural Resources Policy Division. *A Legislative History of the Superfund Amendment and Reauthorization Act of 1986. Public Law 99–499*. Washington, D.C.: U.S. Government Printing Office, 1990.

Farr, William. "Going to Buffalo: Indian Hunting Migrations across the Rocky Mountains. Part 1, Making Meat and Taking Robes." *Montana: The Magazine of Western History* 53, no. 4 (Winter 2003): 2–21.

———. "Going to Buffalo: Indian Hunting Migrations across the Rocky Mountains. Part 2, Civilian Permits, Army Escorts." *Montana: The Magazine of Western History* 54, no. 1 (Spring 2004): 26–43.

Fiege, Mark. *Irrigated Eden: The Making of an Agricultural Landscape in the American West.* Seattle: University of Washington Press, 2000.

Fleischer, Lisa. *Health Study of Milltown, Montana.* Missoula, Mont.: MontPIRG, 1983.

Flores, Dan. *The Natural West: Environmental History in the Great Plains and Rocky Mountains.* Norman: University of Oklahoma Press, 2001.

Forman, Dave. *Rewilding North America: A Vision for Conservation in the 21st Century.* Washington, D.C.: Island Press, 2004.

Gibbs, Lois Marie. *Love Canal and the Birth of the Environmental Health Movement.* Washington, D.C.: Island Press, 2011.

Glasscock, C. B. *The War of the Copper Kings: Greed, Power, and Politics; The Billion-Dollar Battle for Butte, Montana, the Richest Hill on Earth.* Helena, Mont.: Riverbend, 2002.

Gorby, Michael S. "Arsenic in Human Medicine." In *Arsenic in the Environment, Part 2: Human Health and Ecosystem Effects,* edited by Jerome O. Nriagu. New York: John Wiley and Sons, 1994.

Grossman, Elizabeth. *Watershed: The Undamming of America.* New York: Counterpoint, 2002.

Gupta, Shreekant, George Van Houtven, and Maureen Cropper. "Paying for Permanence: An Economic Analysis of EPA's Cleanup Decisions at Superfund Sites." *Rand Journal of Economics* 27, no. 3 (Autumn 1996): 563–82.

Hawley, Steven. *Recovering a Lost River: Removing Dams, Rewilding Salmon, Revitalizing Communities.* Boston: Beacon Press, 2012.

Hays, Samuel P. *Beauty, Health, and Permanence: Environmental Politics in the United States, 1955–1985.* Cambridge: Cambridge University Press, 1987.

Holton, G. D., and H. E. Johnson. *A Field Guide to Montana Fishes.* 3rd ed. Helena: Montana Fish, Wildlife and Parks, 2003.

Jordan, William R., III, and George M. Lubick. *Making Nature Whole: A History of Ecological Restoration.* Washington, D.C.: Island Press, 2011.

Kirk, Andrew G. *Counterculture Green: The Whole Earth Catalog and American Environmentalism.* Lawrence: University Press of Kansas, 2007.

Klyza, Christopher McGrory, and Bill McKibbin, eds. *Wilderness Comes Home: Rewilding the Northeast.* Middlebury, Vt.: University Press of New England, 2001.

Krech, Shepard, III. *The Ecological Indian: Myth and History.* New York: W. W. Norton, 1999.

Lave, Rebecca. *Fields and Streams: Stream Restoration, Neoliberalism, and the Future of Environmental Science.* Athens: University of Georgia Press, 2012.

Lindstrom, Matthew J., and Zachary A. Smith. *The National Environmental Policy Act: Judicial Misconstruction, Legislative Indifference, and Executive Neglect.* College Station: Texas A&M University Press, 2001.

Lowry, William R. *Dam Politics: Restoring America's Rivers.* Washington, D.C.: Georgetown University Press, 2003.

——. *Repairing Paradise: The Restoration of Nature in America's National Parks.* Washington, D.C.: Brookings Institution Press, 2009.

Maclean, Norman. *A River Runs through It.* New York: Simon and Schuster, 1976.

MacMillan, Donald. *Smoke Wars: Anaconda Copper, Montana Air Pollution, and the Courts, 1890–1920.* Helena: Montana Historical Society Press, 2000.

Malone, Michael. *The Battle for Butte: Mining and Politics on the Northern Frontier, 1864–1906.* Helena: Montana Historical Society Press, 1981.

Malone, Michael P., Richard B. Roeder, and William L. Lang. *Montana: A History of Two Centuries.* Rev. ed. Seattle: University of Washington Press, 1976.

Mangam, William Daniel. *Biography of W. A. Clark and His Tarnished Family.* Butte, Mont.: Old Butte, 2007.

Manning, Richard. *Last Stand: Logging, Journalism, and the Case for Humility.* Salt Lake City: Peregrine Smith Books, 1991.

——. *One Round River: The Curse of Gold and the Fight for the Big Blackfoot.* New York: Henry Holt, 1997.

Mead, Daniel W. *Water Power Engineering: The Theory, Investigation and Development of Water Powers.* New York: McGraw-Hill, 1920.

Montana Constitutional Convention, 1971–1972. Helena: Montana Legislature, 1979–1982.

Neuman, Dennis R. "Remediation of the Milltown Dam Sediments in Montana." Presentation at 2014 annual conference, Mine Design, Operations and Closure. http://www.mtech.edu/mwtp/conference/2014_presentations/dennis-neuman.pdf.

Nicholas, Liza, Elaine M. Bapis, and Thomas J. Harvey, eds. *Imagining the Big Open: Nature, Identity, and Play in the New West.* Salt Lake City: University of Utah Press, 2003.

Nickerson, Norma, Christine Oschell, Lee Rademaker, and Robert Dvorak. *Montana's Outfitting Industry: Economic Impact and Industry-Client Analysis.* Missoula: Institute for Tourism and Recreation Research, University of Montana, 2007.

Nriagu, Jerome O., ed. *Arsenic in the Environment, Part 1: Cycling and Characterization.* New York: John Wiley and Sons, 1994.

——. *Arsenic in the Environment, Part 2: Human Health and Ecosystem Effects.* New York: John Wiley and Sons, 1994.

Palmer, Tim. *Endangered Rivers and the Conservation Movement.* Lanham, Md.: Rowman and Littlefield, 2004.

Petersen, Shannon. *Acting for Endangered Species: The Statutory Ark.* Lawrence: University Press of Kansas, 2002.

Power, Thomas Michael. *Lost Landscapes and Failed Economies: The Search for a Value of Place.* Washington, D.C.: Island Press, 1996.

Price, Jenny. "Remaking American Environmentalism: On the Banks of the L.A. River." *Environmental History* 13 (July 2008).

Probst, Katherine N., Don Fullerton, Robert E. Litan, and Paul R. Portney. *Footing the Bill for Superfund Cleanups.* Washington, D.C.: Brookings Institution and Resources for the Future, 1995.

Punke, Michael. *Fire and Brimstone: The North Butte Mining Disaster of 1917*. New York: Hyperion, 2006.

Quammen, David. "Alias, Benowitz Shoe Repair: Heavy Metal, High Water, and a Man in a Neoprene Suit." *Outside*, December 1983.

Quivik, Fred. "Smoke and Tailings: An Environmental History of Copper Smelting Technologies in Montana, 1880–1930." PhD diss., University of Pennsylvania, 1998.

Rainbolt, Jo. *Missoula Valley History*. Dallas: Curtis Media, 1991.

Reagan, Ronald. *The Reagan Diaries*. Edited by Douglas Brinkley. New York: Harper-Collins, 2007.

Reilly, William K. *A Management Review of the Superfund Program*. Washington, D.C.: U.S. Environmental Protection Agency, 1989.

Revesz, Richard L., and Richard B. Steward, eds. *Analyzing Superfund: Economics, Science, and Law*. Washington, D.C.: Resources for the Future, 1995.

Rothman, Hal K. *The Greening of a Nation: Environmentalism in the United States since 1945*. Orlando: Harcourt, Brace, 1998.

Sale, Kirkpatrick. *The Green Revolution: The American Environmental Movement, 1962–1992*. New York: Hill and Wang, 1993.

Salish–Pend d'Oreille Culture Committee and Elders Cultural Advisory Council, Confederated Salish and Kootenai Tribes. *The Salish People and the Lewis and Clark Expedition*. Lincoln: University of Nebraska Press, 2005.

Scott, Ruth Boydston. *Missoula: Trading Post to Metropolis*. Missoula, Mont.: R. Scott, 1997.

Smith, Norman. *A History of Dams*. London: Chaucer Press, 1971.

Stiller, David. *Wounding the West: Montana, Mining, and the Environment*. Lincoln: University of Nebraska Press, 2000.

Stone-Manning, Tracy, and Emily Miller, eds. *The River We Carry with Us: Two Centuries of Writing from the Clark Fork Basin*. Livingston, Mont.: Clark City Press, 2002.

Swanberg, Tim. "The Movement and Habitat Use of Fluvial Bull Trout in the Upper Clark Fork River Drainage." Master's thesis, University of Montana, 1996.

Tyer, Brad. *Opportunity, Montana: Big Copper, Bad Water, and the Burial of an American Landscape*. Boston: Beacon Press Books, 2013.

Vig, Norman J., and Michael E. Kraft. *Environmental Policy in the 1980s: Reagan's New Agenda*. Washington, D.C.: Congressional Quarterly, 1984.

Warren, Louis S., ed. *American Environmental History*. Malden, Mass.: Blackwell Publishing, 2003.

Weisel, George F. "Ten Animal Myths of the Flathead Indians." *Anthropology and Sociology Papers* 18 (1959).

Welch, Alan H., and Kenneth G. Stollenwerk, eds. *Arsenic in Ground Water: Geochemistry and Occurrence*. Norwell, Mass.: Kluwer Academic Publishers, 2003.

Wellock, Thomas R. *Preserving the Nation: The Conservation and Environmental Movements, 1870–2000*. Wheeling, Ill.: Harlan Davidson, 2007.

White, Richard. *Roots of Dependency: Subsistence, Environment, and Social Change among the Choctaws, Pawnees, and Navajos.* Lincoln: University of Nebraska Press, 1983.

U.S. Environmental Protection Agency. *Integrating Water and Waste Program to Restore Watersheds: A Guide for Federal and State Project Managers.* Washington D.C.: Office of Water and Office of Solid Waste and Emergency Response, U.S. Environmental Protection Agency, 2007.

Periodicals and Newspapers

American Indian Quarterly
Anaconda (Mont.) Standard
Congressional Digest
Daily Missoulian (Missoula, Mont.)
Ecological Applications
Engineering News-Record, vols. 52–62, 1904–1909
EPA Journal
Environment
Environmental History
Harper's Magazine
High Country News
Journal of Law and Economics
Missoulian (Missoula, Mont.)
New York Times
Stevensville (Mont.) Northwest Tribune
Rand Journal of Economics
Seattle Times
State Politics and Policy Quarterly
Washington Post
Waste News

Websites

Alliance for the Wild Rockies. "Key Dates and Events in the History of the Bull Trout Listing." Accessed August 19, 2011. http://www.wildrockiesalliance.org/issues/bull trout/history/bulltrout_chronology.html.

American Academy for Park and Recreation Administration. "Bruce Babbitt." Accessed March 16, 2012. http://www.aapra.org/pugsley-bios/bruce-babbitt.

American Rivers. "The 10th Anniversary of the Removal of Maine's Edwards Dam." Accessed November 13, 2014. http://www.americanrivers.org/initiative/dams/projects /the-10th-anniversary-of-the-removal-of-mainea%C2%89uas-edwards-dam/.

——. "Dam Nation: A Snapshot of River Restoration Efforts in the US." Accessed February 3, 2012. https://web.archive.org/web/20120615180010/http://www.american rivers.org/our-work/restoring-rivers/dams/patagonia/dam-removal-map.html.

———. "River Rebirth: Removing Edwards Dam on Maine's Kennebec River." Accessed October 3, 2014. http://www.americanrivers.org/initiative/dams/projects /river-rebirth-removing-edwards-dam-on-maines-kennebec-river/.

Blackfoot Challenge. Home page. Last modified October 3, 2014. http://blackfoot challenge.org/Articles/.

Clark Fork River Technical Assistance Committee. "The Three R's of the Milltown Reservoir Superfund Project." Accessed April 4, 2012. https://web.archive.org /web/20140219202853/http://www.cfrtac.org/061009b.html.

climate-zone.com. "Missoula." Accessed March 10, 2013. http://www.climate-zone .com/climate/united-states/montana/missoula/.

Confederated Salish and Kootenai Tribes. "Jocko River Restoration Project." Accessed March 14, 2012. http://jockoriver.net/Jindex.lasso.

———. "Jocko River Restoration Project: Jocko River Master Plan." Accessed March 14, 2012. http://jockoriver.net/j_master.lasso?page=MasterPlan&side=mp&subside =prjdsc&-session=JRR:3F99556B07ab41608CqtI3540390.

Cornell University Law School. Legal Information Institute. http://www.law.cornell .edu/uscode/text/42/9601.

FundingUniverse. "Atlantic Richfield Company History." Accessed March 10, 2012. http://www.fundinguniverse.com/company-histories/atlantic-richfield-company -company-history.html.

Make It Missoula. "Missoula Nonprofits." Accessed January 18, 2011. http://www .makeitmissoula.com/community/non-profits/.

Missoula Montana. "Open Space." City of Missoula. Accessed April 2, 2012. http:// www.ci.missoula.mt.us/index.aspx?NID=185.

"Missoula's Open Space Program: Making Missoula a Better Place." Accessed April 2, 2012. http://www.ci.missoula.mt.us/DocumentView.aspx?DID=650.

Montana Department of Justice, Montana Lands. "1999 Settlement—Montana v. ARCO." Accessed October 5, 2011. https://dojmt.gov/wp-content/uploads/2011/06 /settlementagreement01.pdf.

National Archives, Federal Register. "Executive Order 12580—Superfund Implementation." Accessed November 15, 2014. http://www.archives.gov/federal-register /codification/executive-order/12580.htm.

Natural Resources Conservation Service. Bob Batz and Raissa Marks. *Bull Trout: Salvelinus confluentus.* Fish and Wildlife Habitat Management Leaflet 36 (January 2006). Accessed August 19, 2011. www.wildlifehc.org/new/wp-content/uploads /2010/10/Bull-Trout.pdf.

Natural Resources Council of Maine. "Edwards Dam and Kennebec Restoration. Accessed April 10, 2012. http://www.nrcm.org/projects-hot-issues/healthy-waters /edwards-dam-and-kennebec-restoration/.

Neuman, Dennis R., KC Harvey Environmental LLC. "Remediation of the Milltown Dam Sediments in Montana." Annual conference, Mine Design, Operations and Closure, 2014. Accessed August 14, 2014. http://www.mtech.edu/mwtp/conference /2014_presentations/dennis-neuman.pdf.

Property and Environment Research Center. Ashley Fingarson. "Cloud with a Copper Lining—From Superfund Site to Golf Course." Accessed March 11, 2012. http://www.perc.org/articles/article220.php.

PRX. "Germaine White, Bull Trout's Gift." Radio interview, KUFM, Montana Public Radio. Accessed March 21, 2012. http://www.prx.org/pieces/63782-germaine-white -bull-trout-s-gift.

SimplyHired. Job search, environmental restoration. Accessed March 15, 2012. http://www.simplyhired.com/a/jobs/list/q-environmental+restoration.

Society for Ecological Restoration. "Mission and Vision." Accessed March 15, 2012. http://www.ser.org/about/mission-vision.

TwoRiversHistory. Home page. Accessed October 2, 2014. http://www.tworivershistory .net/.

U.S. Army Corps of Engineers. "Overview of Dam Decommissioning." Accessed October 12, 2011. https://web.archive.org/web/20090110004834/http://www.crrel.usace .army.mil/sid/Dam_decom/overview.htm.

U.S. Census Bureau. "State and County QuickFacts: Missoula (city), Montana." Accessed January 18, 2011. http://quickfacts.census.gov/qfd/states/30/3050200.html.

U.S. Department of Commerce, National Oceanic and Atmospheric Administration, National Weather Service. "Fast Climate Facts: Missoula, Montana." Accessed January 18, 2011. http://www.wrh.noaa.gov/mso/climfacts.php.

U.S. Environmental Protection Agency. "Cleanup Enforcement: Superfund Unilateral Orders." Accessed June 19, 2012. https://web.archive.org/web/20130703234957/ http://www.epa.gov/compliance/cleanup/superfund/orders.html.

———. "Cleanup Enforcement: Types of Superfund Settlements." Accessed June 19, 2012. https://web.archive.org/web/20121202085709/http://www.epa.gov/compliance /cleanup/superfund/neg-type.html#aas.

———. "Cleanups in My Community." Accessed April 4, 2012. http://iaspub.epa.gov /apex/cimc/f?p=255:63:1603772554611628.

———. "National Priorities List (NPL): Federal Register Notices for NPL Updates." Accessed May 15, 2014. http://www.epa.gov/superfund/sites/npl/frlist.htm.

———. National Service Center for Environmental Publications. Accessed 11/15/2014. http://nepis.epa.gov/.

———. "NPL Site Narrative for Milltown Reservoir Sediments. Accessed May 15, 2014. http://www.epa.gov/superfund/sites/nplsnl/n0800445.pdf.

———. "Region 8: Anaconda Co. Smelter." Accessed July 9, 2014. http://www2.epa .gov/region8/anaconda-co-smelter.

———. "Superfund." Accessed May 21, 2013. http://www.epa.gov/superfund/index .htm.

———. "Superfund: Basic Information." Accessed April 19, 2013. http://www.epa.gov /superfund/about.htm.

———. "Superfund: CERCLA Overview." Accessed May 15, 2014. http://www.epa .gov/superfund/policy/cercla.htm.

———. "Superfund: Natural Resource Damages; A Primer." Accessed May 15, 2014. http://www.epa.gov/superfund/programs/nrd/primer.htm.

———. "Superfund: SARA Overview." Accessed May 15, 2014. http://www.epa.gov /superfund/policy/sara.htm.

———. "Superfund: Superfund and Green Remediation." Accessed May 15, 2014. http://www.epa.gov/superfund/greenremediation/.

———. "Superfund: Superfund Redevelopment Initiative. Accessed May 15, 2014. http://www.epa.gov/superfund/programs/recycle/.

———. "Water: Adopt Your Watershed." Accessed April 29, 2014. http://water.epa.gov /action/adopt/index.cfm.

U.S. Fish and Wildlife Service. "Bull Trout: About Bull Trout: *Salvelinus confluentus.*" Accessed 11/16/14. http://www.fws.gov/pacific/bulltrout/About.html.

———. "Species Profile: Bull Trout (*Salvelinus confluentus*)." Accessed October 2, 2014. http://ecos.fws.gov/speciesProfile/profile/speciesProfile.action?spcode=E065.

U.S. Geological Survey. "USGS Water Resources: USGS Surface-Water Monthly Statistics for Montana." Accessed October 2, 2014. http://waterdata.usgs.gov/mt/nwis /monthly/.

Index

Page numbers in *italics* indicate photographs.